SURPRISE-INSIDE

SURPRISE-INSIDE

amazing
cakes *for*
every occasion—with
a little something
extra inside

CAKES

WITHDRAWN

amanda rettke

WM WILLIAM MORROW *An Imprint of* HarperCollins*Publishers*

Photographs by Susan Powers Photography

HarperCollins books may be purchased for educational, business, or sales promotional use. For information please e-mail the Special Markets Department at SPsales@harpercollins.com.

FIRST EDITION

Designed by Lorie Pagnozzi

Library of Congress Cataloging-in-Publication Data

Rettke, Amanda
    Surprise-inside cakes : amazing cakes for every occasion—
with a little something extra inside / Amanda Rettke.—First edition.
        p.   cm.
    Includes index.
    ISBN 978-0-06-219531-9
    1. Cake.   2. Icings (Confectionery).   I. Title.
    TX771.A47   2013
    641.86′53—dc23   2013034285

14   15   16   17   18   ID6/RRD   10   9   8   7   6   5   4   3   2   1

contents

introduction

I'm proud of being born and raised in Fargo, North Dakota, the biggest small city I know. While most folks know Fargo from the Coen brothers movie, I know it as something entirely different. Kind. Exciting. Friendly. Inspiring. My family still lives there, and we visit as often as life allows. Nothing comforts me more than seeing my old hometown through the eyes of my children, seeing them excited by the things that used to delight me.

Up until the last few years I was purely a consumer of baked goods. I did not make my own cakes or cookies or delectable treats. I didn't even know how! But I did know that in Fargo, Catering by Concordia College has the most amazing banana bread, and Great Harvest Bread Company has the best giant oatmeal cookies, and Quality Bakery makes the best cake donut known to man.

When we moved from a bustling city to the smallest town in Minnesota (well, maybe it's not the smallest, but the population is around 1,000), I didn't know what to do with myself. There was no coffee shop, bakery, or grocery store just around the corner. I couldn't continue to just buy what I wanted—I had to start learning to make things for myself. I was pretty clueless, just getting by in the kitchen.

After I had my first baby in 2005, I started a blog called *i am mommy*. Three people read that blog, and one of them was my mother. Today that blog is one of my most sacred places to share and talk about the amazing little people in my life. At the end of the day, "mommy" is the best description of who I am.

But being a mommy means you must possess a certain amount of know-how. In 2009 I started baking to keep up with the constant flow of birthday parties, Christmas cookie swaps, and church potlucks. I started with cookies—lots and lots of sugar cookies that I decorated with intricate designs and patterns. Having no formal training or experience in the kitchen meant that I needed to teach myself. I watched lots of Food Network and Martha Stewart. I let special occasions, my kids, and my dreams inspire me. My mommy blog was being overrun by baked goods, so I created another blog called *i am baker*. But I kept having babies (five in all!), and the amount of time I was able to dedicate to the intricate art of decorating sugar cookies became less and less. So instead of cookies I started to make cakes. This was pretty strange, because until that point I'd never looked at a cookbook or food blog about cake. In fact I knew nothing about cakes except that we liked to eat them.

When I decided to put a surprise on the inside of my cake for the first time, it seemed like a normal thing to do, simply because I didn't know any better.

The first surprise-inside cake was the Jack-o'-lantern Cake (page 189). I wanted to bring a cake to a church potluck but couldn't think of anything that wasn't too ghoulish or scary, so I settled on a jack-o'-lantern. But then I was nervous that there would be a bunch of pumpkin cakes, so I wanted to try and make mine different. Then it hit me . . . why not put a candle made of cake inside? I pondered that cake for weeks and weeks. My end design was a far cry from the original plan of execution, but every single moment of labor and planning was worth it when I saw people's faces. At that time (2009) there were no surprise-inside cakes—no one had heard of such a thing. I had no idea what to even call it!

Fast-forward to today, and while I may have a little gift for unique baking ideas, I'm not an expert by any means. I'm in awe of all my amazingly talented food-blogging friends, and as an ever-evolving work in progress, I try to learn from the things they share. In the same spirit, I'm happy to share with you here some of the cake designs that have been floating around my mind, and deeply grateful to you for giving me the opportunity to do so.

how to use this book

I highly recommend starting out simple. If you've never made a surprise-inside cake before, don't start with the Cowboy Boot Cake (page 115). Instead start with the Rainbow Cakes (page 41) or the Opposites Cake (page 97), both of which pack a punch with a minimal amount of effort on your part. The most important thing is to practice, practice, practice! Even if a design doesn't turn out the way you envisioned it, the worst-case scenario is that you'll still have a delicious cake to share!

Here are some general tips to help you achieve success with your surprise-inside cakes.

Read the directions from beginning to end before starting

Very important! An average cake in this book is baked over the course of a couple days, so be sure to give yourself enough time, including freezer time. When reading the directions, determine if there are any steps you can complete ahead of time. For instance, you can make buttercream and cake mixture days in advance.

While the steps and carving can seem daunting, the very best possible way to create these cakes is practice! If you're making a surprise-inside cake for a specific event, consider making a practice cake first. I made the surprise-inside heart cake (from my blog http://iambaker.net/heart-cake-surprise-inside-cake/ and similar to the Rainbow Heart Cake) three times in one month and by the third go-round, my confidence level was very high! Once you create your first surprise-inside cake, any anxiety or nervousness will diminish and you'll even be able to start envisioning your own designs.

Embrace learning

While I will always recommend starting out with an easier cake, truly the best way to learn is to try. I made the Cowboy Boot Cake a few times before determining that it just wasn't going to be perfect. Did it mostly look like a cowboy boot? Yes. Would the person who I was giving it to understand the significance? Yes. Did it have to be perfect to be appreciated? Not at all.

Difficulty Levels

I've given each cake a rating of Easy, Medium, or Challenging. I highly recommend starting out simple. You'll gain confidence and get a rhythm, and you'll be off and running on more difficult cakes!

Easy: Anyone with a passion for baking can attempt these cakes! If you can follow a basic recipe, you'll be able to make one of these—the time invested, number of steps, and amount of deconstruction is minimal.

Medium: These cakes require a bit more concentration and attention to detail. You should be comfortable with basic baking, be ready to commit some time, and be willing to step a bit out of your comfort zone.

Challenging: These cakes are for intermediate to advanced bakers. You should have a good grasp of how three-dimensional design works and feel very comfortable with cake deconstruction and reassembly. It's best to attempt a challenging cake only after you've had success with a surprise-inside cake that has an easy or medium level of difficulty. If you're making one of these for a special event, perhaps do a test cake (it will taste great no matter what!). Once you get your hands into a cake and have a successful completion under your belt, your confidence will soar and the "challenging" aspect won't seem so daunting anymore!

Baking

Choosing a cake recipe

The beauty of these cakes is that you can make them with your favorite basic recipes. While I've provided a few favorites starting on page 19, I'm also a fan of using a box mix as a quick substitute (my favorite brand is Betty Crocker). And trust me—people will be so impressed with your skills that they'll never realize you didn't make it from scratch.

If you do want to use your own tried-and-true recipe, I encourage that as well! It can be gluten-free or sugar-free or dairy-free. You'll have a cake that fits your dietary needs but is still visually stunning!

Flavors are up to you

I often choose cake flavors based on their color. You're welcome to use any flavors that fit your needs and wants!

How to Bake a Level Cake (and How to Fix One That Isn't)

Baking a level cake is one of the easiest and most difficult things in the world. Wait, that makes no sense at all. Is it easy or difficult?

Well, it helps if you have a perfectly balanced and uniform recipe. I tested and retested my Chocolate Cake and White Cake recipes, which are specifically created to be level, stable, sturdy cakes (see Cakes, page 19). If any alterations are made to a recipe—say, for flavor, consistency, or even altitude—the result is often a domed cake.

Here are a few tips to help you get perfect cakes every time!

Before Baking

Get an oven thermometer, and use it!

It's very common for your oven to have hot spots or to be off in temperature. If your oven is not heating evenly, you'll especially see this when cooking layer cakes. One layer could be done and the other still raw in the middle. This frustrating occurrence can easily be avoided! I bought a little oven thermometer at the store for a few dollars. Use it to determine the spot in your oven that has the most accurate temperature reading. Try to place cakes in that spot to ensure even and consistent baking.

Prepare your pans

I prefer to use a nonstick spray for my baking pans. Typically I buy one specifically made for baking.

Try using cake strips

These are water-soaked fabric strips that wrap around the cake pan. They can slow down the heat that reaches the edges of the cake and help with even baking. I've used both purchased cake strips and homemade, and I find that they both work equally well. To make your own, simply cut a towel into 1½-inch-wide strips. Thoroughly soak the towel strips in water prior to using. Place them around the outside of the cake pan and either tie them in place with a knot or insert a pin to fasten.

Don't let cake layers touch one another or the sides of the oven when baking.

Make sure there is at least an inch between cake pans. I prefer to bake all my cakes on the center rack of my oven, approximately 9 inches from the top and 9 inches from the bottom. I can successfully bake four 8-inch layers or three 9-inch layers on the middle rack.

Don't be an impatient baker.

Do not check on the cake before the halfway mark. If your cake needs 18 minutes to bake, you are safe to start checking it around the 13-minute mark.

Test the cake about 5 minutes before it's supposed to be done.

Use a toothpick or cake tester and insert it into the cake. If any wet batter clings to the toothpick, the cake needs to bake more. If the toothpick comes out clean or with minimal crumbs, then the cake is done.

After Baking

If your cake still comes out domed even after you've taken precautions, you'll need to carefully cut off the domed part with a serrated bread knife or a cake leveler. You can also place a clean dish towel over the hot cake and gently press down. This can help lower the dome but typically will not make the cake completely flat.

"Doming" isn't my problem—what do I do if my cake falls in the center?

Specifically for these recipes, that's not always a bad thing. As long as the cake is fully cooked in the center, you can still use it! You'll often remove the center of the cake to make the surprise inside, so it might work to your advantage. If you really do need a level cake, you can always bake a new one and keep the collapsed cake in the freezer. It will work wonderfully for a future project!

What to do with extra cake

If you have leftover cake mixture or whole cakes, I would recommend freezing. But by far the best option is to eat it. Since we're often gifting the cakes we make or sharing them on special occasions, having a little leftover cake can be a big treat to all those who have watched you make your labor of love.

Making the Surprise

Freezing cake

Success in making surprise-inside cakes depends on having them firm enough to cut and carve. This involves freezing the cake elements at various points in the process. It's much easier to work on a chilled cake. You'll be able to get precise cuts and almost eliminate crumbs.

For freezing multiple layers of cake, separate each layer with a sheet of parchment paper. If I'm using those layers within hours of placing in the freezer I will use a sheet of parchment between each layer. If the cakes will be in the freezer for longer than 6 hours I will wrap each layer in plastic wrap then place in a plastic sealable bag.

When freezing your cakes during the building process, follow the recommended time frames. For a cake that has been freezing overnight or for 24 hours, you have about 30 minutes to get all your modifications done before it starts getting too soft and crumbly.

My typical recommendation is to freeze a finished cake for at least 6 hours before serving. I often freeze my cakes for days! It's amazing how quickly they thaw out. For a cake that has been in the freezer for 6 hours, I would recommend serving it right out of the freezer.

If you freeze a finished cake for longer, be sure to thaw it in the fridge and not at room temperature. The most important thing is to just make sure your cake has had a chance to chill and set up before serving. The more intricate the design in the interior of the cake, the longer it should chill.

Cutting and carving

The "surprise" in a surprise-inside cake will often depend on carving out cones, cylinders, channels, valleys, and wells from cakes. To direct the cuts, you'll rely on things like cookie cutters, skewers, kitchen string, rulers—even a glass, bowl, or container lid, if it's the right size. For cutting you'll rely on cookie cutters (again), paring knives, serrated knives, soupspoons, baby spoons—whatever gets you the result you need! Whenever possible, I'll suggest that you use commonly found household items.

It may take a bit of nerve the first time you cut out a cake cone, carve out a channel, or insert a column from one cake into another, but believe me—with every step you'll gain confidence. It's like the best parts of kindergarten, art class, and home ec—and it results in a beautiful edible product. What could be more fun?

Making cake mixture

Anytime I make a reference to a cake mixture, I'm talking about cake that has been crumbled and had frosting added to it for the interior of a surprise-inside cake. I learned the secret of cake mixture from another blogger, Angie Dudley (also known as Bakerella), author of *Cake Pops*. She's a genius, and I'm in constant amazement at her work. Without her techniques, there are so many cakes I would never have been able to do!

You can use whatever flavor frosting you desire in the cake mixtures. I have found that most versions of buttercreams and cream cheese frostings work well. I prefer my cake mixture to have a Play-Doh-like consistency, which can sometimes mean using more frosting than cake. Be sure to test as you go, as this can vary based on personal preference.

For the color cake mixtures, you can go about achieving color in one of three ways (see below). I find that all the methods work equally well. I prefer to use gel food coloring, and my favorite brands are AmeriColor and Ateco.

If you have leftover cake mixture, the first and best option is, of course, cake pops! I highly recommend getting Bakerella's book and utilizing her methods and techniques.

How to tint your cake? Choose one!

- Bake a cake in the color you're trying to achieve, then add white (or a coordinating color) frosting to achieve the cake mixture consistency. For example, add ¼ cup white frosting to a blue 8-inch layer cake to achieve a blue cake mixture.

- Add your desired color frosting to crumbled cake. For example, add ¼ cup red frosting to a white crumbled 8-inch layer cake to achieve a red cake mixture.

- Combine crumbled white cake, white frosting, and the desired gel food coloring. For example, add ¼ cup white frosting to a white crumbled 8-inch layer cake. Add 2 to 3 drops of red gel food coloring to achieve a pink cake mixture.

Frosting and Decorating

I love frosting cakes. To me, they're a blank canvas ready to be beautified! But getting the perfectly crisp, clean base cake isn't always easy. Through much trial and error, I've found a method that works for me, and I believe it can work for you as well.

feel free to mix and match

If you see a frosting technique on one cake and want to pair it with a different cake, you should go for it! Just make sure you have all the necessary ingredients, tools, and gel food coloring and/or sprinkles needed to achieve your desired look.

cake decorating tools

Rotating cake stand—useful for making many of the surprises in this book and really handy for frosting and decorating as well

Baker's blade or mudding trowel—handy for lifting and moving cakes

Offset spatula—great for frosting, whether your goal is smooth lines or swirls and different effects

Disposable plastic pastry bags and decorating tips—great to have lots on hand for experimenting!

Coupler set—this makes it easy to swap in different bags of frosting, which is handy when you're frosting a large area or switching out different tips.

1. Place a leveled cake layer on a rotating cake stand. (You can also place your cake on your display cake plate and put that cake plate right on your rotating cake stand.)

2. Using an offset spatula, place approximately ½ cup frosting on the center of the cake. The amount can vary depending on the size of your cake.

3. Place the spatula just above the surface of the cake and hold it still. Using your free hand, rotate the cake stand. You do not need to move the spatula, only the stand. Continue until you have a nice even surface of frosting anywhere from ¼ to ½ inch thick. The frosting will spill out over the edges, and that's okay.

4. Place the final layer of cake on top. If the cake has more layers, just repeat steps 3 and 4 until you reach the top of the cake.

5. Now you'll create the crumb coat of the cake—a thin layer of frosting that seals in all the crumbs to allow the next frosting layer to go on cleanly. Holding the offset spatula firmly against the side of the cake, begin to turn the cake stand again, spreading the frosting that has spilled out between the layers smoothly over the surface. Add more frosting to the spatula as needed—just keep the layer thin and smooth.

The crumb coat creates a nice, clean, smooth working surface for your cake decorations.

6. Place approximately ½ cup of frosting on top of the cake. Place the offset spatula level on the top surface of the cake and rotate the cake stand, creating a smooth, flat surface.

7. Now place the spatula firmly against the side of the cake. Rotate the cake stand and smooth out the excess frosting. If you need to wipe excess frosting off your spatula, be sure to do this into a separate bowl so you don't get any crumbs in your frosting. If you come to a spot that needs more frosting, take some from the excess frosting bowl and continue to smooth the edges.

8. When you're happy with the smoothness, turn your attention back to the top. Take your spatula and hold it just outside the edge of the cake.

9. Push the spatula toward the center of the cake, dragging along any excess frosting. Then lift the spatula and wipe it clean. Repeat around the entire perimeter of the cake.

10. Place the cake on the cake stand in the freezer. You want the cake to be very firm so it can easily be moved from the stand to a display plate, where it will receive the final decoration. This usually takes at least 2 hours.

11. When your cake is nice and frozen, slide a baker's blade under the cake as far as you can. Gently lift the whole cake, then quickly transfer it to your display plate, center the cake, and remove the baker's blade. Place the cake back onto a rotating cake stand. (I switched to a lower stand to make things easier.)

12. Drop approximately ½ cup of frosting in the center of the cake and use the same method as in step 3 to create a level top.

13. I switch between a baker's blade and an offset spatula for the sides of the cake, using whatever tool feels most comfortable. Add more frosting around the base as needed. As in step 5, hold the spatula squarely against the side of the cake and rotate the cake stand, removing any excess frosting as you go. If there are spots that appear too thin, just go back and add a bit more frosting.

The very best way to get smooth frosting on a cake is to practice. I'm still learning new techniques and methods all the time!

I've found that while I love the look of sharp, clean lines on a buttercream-frosted cake, perfection is not necessary. The less time I spend worrying about how crisp my edges are, the more time I can focus on creating a beautiful, fun design.

The perfect blank canvas.

How to Serve a Surprise-Inside Cake

Many cakes have special shapes in the center that require carving and/or cutting out the interior. This can change the structural integrity of the cake, so you'll want to be careful when serving. I always try to serve a surprise-inside cake cold and place it back in the fridge after everyone has gotten a piece to keep it chilled for a second round. Most cakes will be just fine if kept at room temperature as well—you'll just want to gauge the integrity after you've cut into it.

Basic Cakes and Frostings

I'm happy to share some of my tried-and-true recipes with you—cakes and frostings that you can use as the building blocks for my surprise-inside cakes. A few of these recipes were handed down to me by Great-Great-Grandma Inga from her *Hope Lutheran Ladies Aid Cook Book* from 1955.

The instructions in that beautiful cookbook are quite brief, so I've rewritten them in my own voice while still honoring the classic simplicity of the great bakers of my past.

I've also included the recipes for my favorite frostings. My basic buttercream recipe yields about 4 cups of frosting. I tend to have a heavy hand when decorating, so I tried not to get too specific with my suggested quantities. The general ratio goes a little like this: Figure about ½ cup of frosting between each layer of cake and 1 cup of frosting to crumb-coat one layer of cake. Any decorations will mean additional frosting is needed, but all exterior decorations are optional, of course.

cakes

frostings

White Cake

My goal for this cake was the perfect crumb. I wanted a cake that was sturdy, moist, but easy to work with when carved and manipulated. The final result is all those things *and* delicious!

3 cups cake flour

½ teaspoon salt

1 tablespoon baking powder

1 cup (2 sticks) unsalted butter, at room temperature

2 cups sugar

1 cup milk, at room temperature

1¼ teaspoons clear vanilla extract

¼ teaspoon almond extract

5 large egg whites, at room temperature

1. Set a rack in the center of the oven, then preheat the oven to 350°F. Prepare two 8-inch round cake pans as directed on page 8.

2. Sift together the cake flour, salt, and baking powder into a large bowl and set it aside.

3. In a standing mixer using the paddle attachment or in a large bowl, cream the butter and sugar on medium speed for about 2 minutes, or until light and fluffy.

4. In a large measuring cup, combine the milk, vanilla, and almond.

5. Add one-third of the flour mixture to the butter mixture and blend on a low speed for about 30 seconds. Add one-half of the milk mixture and blend until just combined. Add in half of the remaining flour mixture and blend until just combined. Add the remaining milk mixture, blend for a few seconds, then the remaining flour mixture and blend at low speed until the ingredients are just incorporated.

6. Pour the batter into a large bowl and clean the stand mixer bowl. Switch in the whisk attachment.

7. Add the egg whites and beat them on medium to medium-high speed to firm peak stage. (When the whisk is held sideways, the peaks will hold and the ridges will be distinct. It's okay if the tips of the peaks fold back on themselves.) Then gently fold the egg whites into the cake batter.

8. Transfer the batter to the prepared cake pans. Place the cakes on the center rack of the oven and bake them for 20 to 26 minutes, or until a toothpick inserted into the center of the cake comes out clean. Let the pans cool for 5 to 10 minutes. Carefully remove cakes and set them on wire racks until they reach room temperature.

Chocolate Cake

Two years ago I would have balked at a chocolate cake recipe that included sour cream. After testing dozens and dozens of recipes, I found that this one was the best—the best flavor, the best texture, the best overall chocolate cake. If you want more depth of flavor you can add a good-quality semisweet or bittersweet chocolate!

¾ cup unsweetened cocoa

1½ cups sugar

1½ cups cake flour

1 teaspoon baking soda

Pinch of salt

2 large eggs, at room temperature

1 cup sour cream, at room temperature

2 teaspoons good-quality vanilla extract

½ cup hot coffee or hot water

1 cup milk chocolate (about 4 ounces), finely chopped

1. Set a rack in the center of the oven, then preheat the oven to 350°F. Prepare two 8-inch round cake pans as directed on page 8.

2. In a standing mixer, using a paddle attachment, blend the cocoa, sugar, cake flour, baking soda, and salt on a low speed for 30 seconds.

3. Add the eggs, sour cream, and vanilla and mix for 1 minute on medium-low speed.

4. Remove the bowl from the mixer and add the coffee and chocolate. (Be sure to finely chop the chocolate, as large chunks may not melt during baking.) Stir by hand until the ingredients are fully incorporated.

5. Pour the batter into the prepared pans and bake the cakes for 28 to 32 minutes, or until a toothpick inserted into the center of the cakes comes out clean. Let the pans cool for 5 to 10 minutes. Carefully remove the cakes and set them on wire racks until they reach room temperature.

dark chocolate cake

To make this recipe a dark chocolate cake, substitute bittersweet (at least 60 percent cocoa) for the milk chocolate.

Red Velvet Cake

Even though you typically see Red Velvet cake around holidays, I can't help but love it all year round. It's such a beautiful and classic dessert!

2 cups all-purpose flour

1 teaspoon salt

1 teaspoon baking soda

1½ cups sugar

½ cup shortening

2 large eggs, at room temperature

4 tablespoons unsweetened cocoa

1 teaspoon red gel food coloring

2 teaspoons good-quality vanilla extract

2 tablespoons hot coffee

1 cup buttermilk, at room temperature

1. Set a rack in the center of the oven, then preheat the oven to 350°F. Prepare two 8-inch cake pans as directed on page 8.

2. Sift the flour, salt, and baking soda into a medium bowl and set it aside.

3. In a standing mixer using the paddle attachment or in a large bowl, cream the sugar and shortening on medium speed for 3 to 4 minutes, or until creamy.

4. In a separate bowl, beat the eggs until light and fluffy, about 3 minutes. Slowly add the eggs to the sugar mixture, mixing as you go.

5. In a small bowl, combine the cocoa, food coloring, vanilla, and hot coffee. Immediately add the cocoa mixture to the sugar mixture and mix until fully combined.

6. Add one-third of the flour mixture to the sugar mixture and mix to combine, followed by half the buttermilk. Add half the remaining flour mixture, the rest of the buttermilk, and the rest of the flour mixture, mixing after each addition.

7. Pour the batter into the prepared cake pans and bake for 25 to 35 minutes, or until a toothpick inserted into the center of the cakes comes out clean. Let pans cool for 5 to 10 minutes. Carefully remove cakes and set them on wire racks until they reach room temperature.

Note: *If you do not have any buttermilk on hand, there is an easy trick you can use! Place a tablespoon of white vinegar in a 1-cup (or larger) liquid measuring cup, then add enough milk (I prefer using 2% or whole milk) to bring the liquid up to the 1-cup line. Let stand for 5 minutes, then use as much as your recipe calls for.*

Strawberry Cake

This is a big cake. A big cake with flavor! I actually prefer making this recipe in three pans. And being the sucker for layer cakes that I am, I love that it is three perfect layers of strawberry bliss. Every frosting complements it—chocolate, vanilla, cream cheese—but my favorite topping is whipped cream!

4 large eggs, separated, at room temperature

2 cups sugar

1 cup (2 sticks) unsalted butter, at room temperature

1 tablespoon good-quality vanilla extract

½ teaspoon almond extract

3 cups plus 2 tablespoons flour

½ teaspoon salt

2 teaspoons baking powder

⅔ cup strawberry puree (blend 1 cup thawed frozen berries and 1 or 2 tablespoons water)

2 cups fresh strawberries, diced

1. Set a rack in the center of the oven, then preheat the oven to 350°F. Prepare three 8-inch round cake pans as directed on page 8.

2. In a standing mixer, using the whisk attachment, beat the egg whites until soft peaks form, about 3 minutes. Add 1 cup of sugar, 1 tablespoon at a time, whipping to maintain soft peaks. Transfer the egg white mixture into a large bowl (it will be about 5 cups in volume) and set it aside. Clean the mixer bowl.

3. Change to the paddle attachment and use the mixer to cream the butter and 1 cup of sugar at medium speed until fluffy, about 2 minutes. Add the vanilla and almond and mix to combine. Add the egg yolks one at a time, beating after each addition.

4. Sift together 3 cups of the flour, the salt, and the baking powder into a medium bowl.

5. Add one-third of the flour mixture to the batter and mix to combine, followed by half of the strawberry puree, half the remaining flour mixture, the remaining strawberry puree, and the remaining flour mixture, mixing after each addition. The batter will be thick.

6. Using a spatula, fold in the egg whites.

7. In a large bowl, coat the diced strawberries with the remaining 2 tablespoons of flour. Using a spatula, gently fold them into the batter.

8. Pour the batter into the prepared pans and bake for 35 to 40 minutes, or until a toothpick inserted into the center of the cake comes out clean. (If you use two 8-inch cake pans, bake for 45 to 50 minutes.) Let pans cool for 5 to 10 minutes. Carefully remove the cakes and set them on wire racks until they reach room temperature.

Decadent Brownies

I have to admit, one of my favorite cakes in the whole book is the Neapolitan Hi-Hat Cake (page 93). The rich brownie covered in fresh strawberry cake topped with mounds of whipped cream is as close to cake perfection as I can imagine. While this recipe is perfect for that cake, these brownies are fabulous enough to stand alone!

1 cup granulated sugar

½ cup packed light brown sugar

½ cup (1 stick) unsalted butter, at room temperature

2 teaspoons good-quality vanilla extract

2 large eggs, at room temperature

½ cup all-purpose flour

¼ cup cake flour

½ cup unsweetened cocoa

¼ teaspoon salt

½ cup milk chocolate (about 2 ounces), finely chopped

1. Set a rack in the center of the oven and preheat the oven to 350°F. Prepare an 8-inch round cake pan as directed on page 8.

2. In a standing mixer, using the paddle attachment, or in a large bowl and using a hand mixer, combine the granulated sugar, brown sugar, butter, and vanilla and mix on medium speed until light and fluffy, about 2 minutes. Beat in the eggs one at a time until well blended.

3. Sift the all-purpose flour, cake flour, cocoa, and salt in a medium bowl. Gradually add the flour mixture to the butter mixture with the mixer on low speed.

4. Remove the bowl from the mixer and fold in the chocolate by hand, using a spatula. Pour the batter into the prepared pan and bake the brownies for 25 to 30 minutes, or until a toothpick inserted into the center comes out mostly clean (a crumb or two is just fine). Let the brownies cool to room temperature in pan.

Notes:
- *You can double this recipe and use a 9 × 13-inch pan (use the same cooking time).*
- *You can use semisweet chocolate in place of the milk chocolate for even more depth of flavor.*

Metric Conversion Chart

⅛ teaspoon = 0.5 mL

¼ teaspoon = 1 mL

½ teaspoon = 2 mL

1 teaspoon = 5 mL

1 tablespoon = 3 teaspoons = ½ fluid ounce = 15 mL

2 tablespoons = ⅛ cup = 1 fluid ounce = 30 mL

4 tablespoons = ¼ cup = 2 fluid ounces = 60 mL

⅓ cup = 3 fluid ounces = 80 mL

½ cup = 4 fluid ounces = 120 mL

⅔ cup = 5 fluid ounces = 160 mL

¾ cup = 6 fluid ounces = 180 mL

1 cup = 8 fluid ounces = 240 mL

350°F = 180°C

basic buttercream

This recipe is a perfect starting point. You can modify it in so many ways . . . using Champagne instead of heavy cream, or all butter and no shortening. Feel free to experiment and make it your own!

½ cup (1 stick) unsalted butter, at room
    temperature
½ cup shortening
2 teaspoons good-quality vanilla extract
Dash of salt
One 2-pound bag confectioners' sugar
    (about 7 cups)
¼ to ½ cup milk or heavy cream

Combine the butter, shortening, vanilla, and salt in a standing mixer using the paddle attachment (or in a large bowl with a hand mixer). Add the confectioners' sugar 1 cup at a time, alternating with the milk or heavy cream, and blend until you have used it all. If the frosting is too thick, add more milk. If it is too thin, add more confectioners' sugar.

Notes:

- *To create a bright-white buttercream, use a clear vanilla extract and 1 cup of shortening—omit the butter.*
- *If using a microwave to achieve room-temperature butter, heat in 10-second increments and watch very closely. It is important for the butter to be at room temperature and not melted.*
- *I've found that this keeps in the fridge for at least three weeks—just be sure to let it warm up to room temperature before using.*

honey buttercream

The perfect frosting for the Bee Cake on page 261! Since honey is such a wonderful natural sweetener, I've decreased the amount of sugar in this recipe versus the basic buttercream.

½ cup (1 stick) butter, at room temperature
½ cup shortening
¼ cup honey
4 to 6 cups confectioners' sugar
3 to 5 tablespoons heavy cream (more if needed)

In a stand mixer using the paddle attachment, or in a large bowl, cream together the butter, shortening, and honey at medium speed until fully incorporated. Turn the mixer to low speed and slowly add the confectioners' sugar (start with 4 cups). Add the heavy cream as needed to create the texture of frosting you prefer and blend for about 3 minutes on medium-high, until light and fluffy. Check the consistency and sweetness and add up to 2 cups more confectioners' sugar, as you prefer. If the frosting gets too stiff, add more heavy cream, 1 tablespoon at a time.

chocolate buttercream

This is one of my favorite frostings of all time. It's also one of those frostings where the quality of your ingredients makes a difference. Opt for a good-quality cocoa powder and vanilla if possible. I often double the recipe so that I have some in the fridge—it's the perfect go-to frosting for any flavor cake! You can see it on the Zebra Cake (page 105).

½ cup (1 stick) unsalted butter, at room temperature
½ cup unsweetened cocoa powder
1 teaspoon good-quality vanilla extract
4 to 5 cups confectioners' sugar
¼ cup milk

In a stand mixer using the paddle attachment, or in a large bowl, cream the butter, cocoa, and vanilla at medium speed for 1 to 2 minutes. Add the confectioners' sugar 1 cup at a time, adding in milk whenever the frosting gets too stiff or the mixer starts to struggle. After 4 cups of sugar have been added, add the rest of the milk and mix until the frosting is smooth and silky (about 2 minutes). For a stiffer consistency, add more confectioners' sugar, ¼ cup at a time. If the frosting gets too stiff, add in more milk, 1 tablespoon at a time.

cream cheese frosting

Just four ingredients and you have one of the richest, creamiest frostings ever. It's perfect for cake, of course, but it's also great on banana bread, pumpkin bread, and spread on top of your favorite cookie! It's wonderful on the Vintage Cake, page 45.

One 8-ounce package cream cheese, at room temperature
½ cup (1 stick) unsalted butter, at room temperature
1 teaspoon good-quality vanilla extract
4 cups confectioners' sugar
¼ cup milk

Combine the cream cheese, butter, and vanilla in a stand mixer, using the paddle attachment, and blend for 1 to 2 minutes on medium speed, or until fully incorporated. Add the confectioners' sugar, 1 cup at a time, and continue mixing until the frosting is light and creamy. For a stiffer consistency, add more confectioners' sugar, ¼ cup at a time. If the frosting gets too stiff, add milk, a tablespoon at a time.

CELEBRATING FAMILY

this chapter is all about celebrating the people and things we hold dear, including our family, friends, pets, country, and the joyous celebration of a new baby.

stripe birthday cake

My little girl Audrey has a December birthday, and I try to make sure her day is all about her ... no red and green, and lots of pink and gold! When I showed the fabric in this picture to my little girl, her eyes lit up and her smile widened and she jumped and danced and said, "I love it, Mama!" The Stripe Birthday Cake was born, inspired by a simple section of cloth. It made a big impact on a little girl who dreams of twirling dresses and grand castles, and the best part is that the fabric turned into a spectacular robe, fit for a royal princess!

2 recipes White Cake,
page 19

1 recipe Basic Buttercream,
page 24

Teal, pink, orange, peach,
purple, and yellow gel
food coloring

SPECIAL EQUIPMENT:
*6 6-inch round cake pans; cake
leveler; small offset spatula;
birthday candles*

DIFFICULTY: *Easy*

Baking

1. Prepare a single recipe of white cake, which will yield about 4 cups of batter. Divide the batter evenly among 4 bowls.

2. Use teal, pink, orange, and peach food coloring to tint the batter in the 4 bowls.

3. Using 6-inch cake pans, bake the 4 tinted layers. Set them aside to cool.

4. Prepare a second recipe of white cake. Divide the batter in two and tint one bowl with purple food coloring. Pour the tinted batter into two 6-inch cake pans, bake for 30 to 36 minutes, and set them aside to cool. (If you want to make additional layers, feel free to tint the remaining batter whatever colors you like.)

Making the Surprise

5. You now have 6 cake layers. Find the shortest cake of all the colors and use it as your height guide. Level all the other layers to the same height (see page 9).

6. Prepare the buttercream and tint half of it yellow.

7. Place a purple layer on a cake stand. Using the offset spatula, cover in about ¼ cup yellow buttercream.

9. Repeat with the pink, orange, and peach layers, then top with the second purple layer.

10. Clean the offset spatula, and use the remaining yellow buttercream to cover the cake in a crumb coat (see page 13). Freeze the cake for 30 minutes.

Frosting and Decorating

11. Cover the cake in a smooth layer of white buttercream (see page 14) and decorate with fun coordinating candles.

8. Place a green layer on top and cover it in about ¼ cup yellow buttercream.

maypole cake

2 recipes White Cake, page 19

2 recipes Basic Buttercream, page 24

Purple, blue, green, yellow, orange, and pink gel food coloring

SPECIAL EQUIPMENT:
4 8-inch round cake pans; long serrated knife or cake leveler; offset spatula; thin ribbons in pastel colors; ⅛-inch wooden dowel (found at craft or home improvement stores), cut to a 24-inch length

DIFFICULTY: *Easy*

I love maypoles. Something about the idea of laughing children dancing around holding ribbons just makes me smile. That and the fact that for some reason they remind me of my favorite movie ever, *Pride and Prejudice*.

There are no maypoles in *Pride and Prejudice*, by the way.

But there's hunky Mr. Darcy.

And long, meaningful glances.

And breathless declarations of undying love.

All *before* the first kiss.

Sigh

What was I talking about again? Oh, yes. Cake.

Baking

1. Bake 4 white cake layers in 8-inch round cake pans. Cool the cakes to room temperature, then freeze them for at least 2 hours.

2. Using a long serrated knife, carefully cut each layer in half horizontally, to create 8 layers. You need only 7 layers to make this cake, so you can use the remainder to create mini cake pops, or just to snack on! (Or—just for fun—create a doll version of the maypole cake by cutting the extra cake into tiny layers with a 2-inch cookie cutter. Use a different color buttercream between each layer!)

3. Prepare 1 recipe buttercream and divide it among 6 bowls (about ¾ cup per bowl).

4. Add 1 or 2 drops of gel food coloring to each bowl and mix.

Making the Surprise

5. Place 1 cake layer onto a cake stand. Using the offset spatula, smooth all of the purple buttercream on top of the layer. Since there will be so much weight on this cake and you want each layer to be the same height, spread the buttercream just to within ½ inch of the edge. Gravity will take care of the rest; this method will keep the frosting from spilling too much over the edge. Clean the spatula.

6. Place another cake layer on top of the purple buttercream and frost it with the blue buttercream. Repeat, frosting in reverse rainbow order (green, yellow, orange, pink) cleaning the spatula between layers until you've used up all the buttercream and topped with the final layer of cake. (If you like, you can freeze the cake for

1 hour after you've added the green buttercream. This helps set the buttercream and works as another level of protection against the force of gravity.)

7. Refrigerate the cake while you prepare the second batch of buttercream.

8. Cover the cake in a crumb coat (see page 13). Refrigerate the cake for at least one hour.

9. To create a textured effect with the buttercream, simply place it onto the cake in dollops to create a thick layer of frosting. Using the back of a small offset spatula, make half-circle motions, gently creating a wave effect.

Frosting and Decorating

10. Wrap a colorful ribbon around the entire length of the dowel. You can secure it with double-sided tape or superglue. Choose a variety of colors of ribbon and secure them to the top of the dowel. Cover the top with a big bow, sticker, or fluffy pom-pom.

11. Insert the dowel into the center of the cake, deep enough that it's stable and secure. Drape the ribbons over the cake.

tips and tricks

Try standing up slices of cake when you serve them for that extra wow factor. Make sure to cut a thick-enough slice so that the cake has some stability, then transfer the slice to a plate with a cake server. This will help keep the slice upright and also avoid fingerprints in the frosting.

ombre cake

If I could use any word I wanted to describe this cake, it would be *multifarious*. It truly has something for everyone…a variety of shapes, textures, and flavors! This rich, elegant cake can work for almost any occasion, but for me it screams masculinity. Perfect for Father's Day or your favorite guy's birthday!

SPECIAL EQUIPMENT: *2 8-inch round cake pans; offset spatula; disposable plastic pastry bags; #127 decorating tip (or #104 if you want smaller petals)*

DIFFICULTY: *Easy*

Baking

1. Bake each cake recipe in two 8-inch cake pans. Let cool. You'll be using only one layer from each of the three cakes, so you can wrap and freeze the other layers for future use.

Making the Surprise

2. Prepare the buttercreams. Place all the white buttercream in a pastry bag, tie it off, and set it aside. Divide the chocolate buttercream between 2 bowls and set one bowl aside. Tint the other bowl with the dark-brown food coloring.

3. Place the dark chocolate layer on a cake stand and spread a thin layer (about ¼ cup to ½ cup) of dark buttercream on top.

4. Add the milk chocolate cake and frost the top with a thin layer (about ¼ cup to ½ cup) of the untinted chocolate buttercream using an offset spatula.

5. Add the white cake layer.

Frosting and Decorating

6. Place the remaining chocolate and dark chocolate buttercreams in pastry bags and cut off the tips. Place the decorating tip into a pastry bag.

7. Starting at the base of the bottom layer, frost the cake with petal shapes. Let the tip do the work for you: Hold the smaller end out and apply pressure, then simply move the tip slightly to the right. Release the pressure and move to the next petal. You'll want the petals to slightly overlap.

8. Working from the bottom, make two full rows of petals across the dark chocolate layer, then switch to the milk chocolate frosting.

9. Using the same technique, do two full rows of milk chocolate frosting on the milk chocolate layer.

10. Now switch to the white frosting and continue around the cake and over the top. This cake can be served immediately or kept in the fridge for up to 3 days.

rainbow cakes

I'm obsessed with rainbows. If you happen to read my blog you already know that, since I talk about them quite often! I've made rainbow cakes, double-rainbow cakes, cupcakes, pancakes—basically anything I can stuff a rainbow into, I will. This cake is the grandbaby of my first rainbow cake. The adaptation is minimal, but the impact is just as grand!

Note: *This recipe makes two 3-layer cakes.*

Baking

1. Prepare the batter for 1 recipe white cake.

2. Divide the batter evenly among 3 bowls. Add 1 to 10 drops of red, orange, and yellow food coloring per bowl, depending on your desired level of color intensity. (I like a vibrant cake, so I used 10 drops of each color.)

3. Bake each layer in a 6-inch cake pan and cool the layers to room temperature.

4. Repeat this process to make a second cake, using the purple, blue, and green food coloring.

5. Find the shortest of the six cakes, then level the others to the same height (see page 9). This is very important!

---

2 recipes White Cake, page 19

Red, orange, yellow, purple, blue, and green gel food coloring

2 recipes Basic Buttercream, page 24

SPECIAL EQUIPMENT: *3 6-inch round cake pans; small offset spatula; 2-inch round cookie cutter; long, thin knife; disposable plastic pastry bags; #21 or #32 decorating tip*

DIFFICULTY: *Easy*

TIP: *Whether you use my recipe or a cake mix, be sure to sift the dry ingredients. Any lumps will end up as little white dots in the brightly colored cake!*

6. Prepare the buttercream. Place ½ cup in a small bowl and mix in enough orange food coloring to match the orange cake layer. Repeat to make yellow, blue, and green buttercream.

Making the Surprise

9. Add the yellow cake layer.

10. Repeat steps 7 through 9 to create a second cake with the purple, blue, and green layers, using the blue and green buttercream and cleaning the spatula between changes of color.

7. Place the red layer on a cake stand and frost the top with a thin layer of the orange buttercream using the small offset spatula. Do not spread the frosting all the way to the edge, just to within ½ inch of the edge. Clean the spatula.

8. Add the orange cake layer and spread the top with a thin layer of the yellow buttercream.

11. Gently push the cookie cutter into the center of one of the cakes to use as a guideline, then use a long, sharp, thin knife to slice all the way through and cut out a column. Repeat this process on the other cake. You're going to place the column from one cake into the other cake and vice versa, so keep the cuts as neat and even as possible. If the cake feels crumbly or soft, freeze it for at least one hour.

12. The easiest way to remove the column is to start with a very chilled cake. Place the cake on its side and gently push on the cylinder with your thumb. Apply pressure and push it through and then out of the cake. Carefully set the cylinder upright. Do this to both cakes. Carefully replace the cylinder with the opposite color scheme into the cake and vice versa.

13. Crumb-coat both cakes with the untinted buttercream (see page 13) and chill.

Frosting and Decorating

14. Using the small offset spatula, cover the cakes in a smooth coat of white buttercream (see page 14). Divide the remaining buttercream equally among 6 bowls and tint it red, orange, yellow, green, blue, and purple. Fill 6 plastic disposable pastry bags with the frosting.

15. Insert the decorating tip into a pastry bag and then add your first bag of frosting. Create a pattern of herringbone stripes on one of the cakes, starting from the top edge of the cake and working your way down the side of the cake. Simply pipe out a scallop at a 45-degree angle, then come back over from the opposite angle.

16. On the other cake, pipe out rings of scallops on the top of the cake in rainbow order, starting at the center.

vintage cake

This cake is inspired by Grandma Audrey, whom I named my little girl after. Grandma is beyond creative. She's written and illustrated children's books, is a master oil and watercolor painter, and is even a greeting card designer! You can see a snippet of her talent in the beautiful drinking glass behind this cake, which she painted for me.

1 box white cake mix
(or 1 recipe White Cake,
page 19)

1 box brownie mix
(or 1 recipe Decadent
Brownies, page 23)

2 squares (2 ounces)
semisweet chocolate,
finely chopped

7 ounces sweetened
condensed milk (half of
a 14-ounce can)

1 recipe Cream Cheese
Frosting, page 25

She's also an inspiration in the kitchen. For years and years I begged her to share this recipe with me. Imagine my surprise when she revealed it was a box mix! With her permission, I'm so delighted to share this Vintage Cake with you. The flavor combination is divine and the brownie layer adds a texture you'll want to use again and again. You can of course use your favorite recipes from scratch if you like.

SPECIAL EQUIPMENT:
*3 9-inch round cake pans;
small offset spatula*

DIFFICULTY: *Easy*

Baking

1. Pour the white cake batter into two 9-inch round cake pans. If the edges start to get too dark, place a ring of aluminum foil around the edges. Cool the cakes to room temperature.

2. Bake the brownies in a 9-inch round cake pan. The level of the batter should not be higher than three-fourths up the side of the pan; bake any extra batter as cupcakes.

3. To make the chocolate sauce, place the chocolate, sweetened condensed milk, and 1 tablespoon of water in a small saucepan and cook over low heat, stirring until the chocolate is melted, about 5 minutes.

Making the Surprise

4. While it's still warm, place the brownie layer on a cake stand. Slowly pour the chocolate sauce over the brownie layer, letting it be absorbed into the brownie. Sometimes I use all the chocolate sauce and allow the sauce to spill over the sides of the brownie and sometimes I use less and contain the chocolate sauce to the top. Allow the sauce to absorb for at least 30 minutes. Before you assemble the rest of the cake, clean the cake stand of any excess sauce.

5. Prepare the cream cheese frosting.

6. Place a white cake layer on top of the brownie and top with approximately ¼ cup of cream cheese frosting using the offset spatula.

7. Add the second white cake layer and cover the cake in a crumb coat of cream cheese frosting (see page 13).

Frosting and Decorating

8. Cover the cake in a layer of cream cheese frosting. To achieve a vintage, rustic look, simply swoop the offset spatula through the frosting in C shapes. Do this over the entire cake.

9. Chill the cake until ready to serve.

apple cake

I totally had a teacher in mind when creating this cake. Which is funny, since I homeschool.

But we all have many teachers in life, whether at school, Sunday school, karate class, or even Walmart, where helpful grandmas insist on demonstrating the proper way to discipline a screaming child in the middle of the frozen food aisle.

I like to consider myself a lifelong learner, and I hope to impart that to my kids as well! I also hope to teach them that it's important to say "thank you" to people who impact our lives. And it's even better when we can do it with cake!

Baking

1. Bake 4 white cake layers in 8-inch round cake pans. Cool to room temperature.

2. Bake the red velvet cake in the pan of your choice, such as a cupcake tin or a 9 × 13-inch pan. Cool to room temperature.

3. Prepare the buttercream. Tint ½ cup leaf green and ½ cup brown. Leave the remaining buttercream white.

4. Place 1 white layer cake on a piece of parchment paper. Using the offset spatula, cover it in a thin layer of white buttercream, then add a second layer. Repeat to make a second 2-layer cake.

---

2 recipes White
Cake, page 19

½ recipe Red Velvet Cake,
page 21

1½ recipes Basic
Buttercream, page 24

Green, red and brown,
gel food coloring

SPECIAL EQUIPMENT:
*4 8-inch round cake pans;
cupcake tin or 9 x 13-inch pan;
offset spatula; 4½-inch round
cookie cutter; small, sharp knife;
disposable plastic pastry bags;
#65 and #4 decorating tips*

DIFFICULTY: *Medium*

5. Freeze the cakes for at least 6 hours, preferably overnight.

Making the Surprise

6. Remove the cakes from the freezer. Place a 4½-inch cookie cutter in the center of one of the cakes and press in slightly to create a guideline. Repeat with the second cake. Then insert a toothpick or skinny skewer into the center of each cake and then remove it. This will be your center guide point.

7. Insert a small, sharp knife 1 to 2 inches deep in the guide point at a 45-degree angle. The knife should be pointing away from the center of the cake. Slowly move the knife in a circle, keeping your depth and angle steady.

8. Insert the knife at the guideline at a 45-degree angle toward the center of the cake. Slowly cut around the cake.

9. Use a spoon to go in and carve out the rounded shape of the bottom of an apple, leaving an "island" in the center, which will be the indentation at the bottom of the apple. The indentation should go all the way through one layer of cake and about halfway through the bottom layer. Do not go all the way through both layers. Place the pieces of cutout cake into a bowl.

10. Repeat with the other cake, this time carving the shape of the top of the apple. This time the "island" represents the well at the top of the apple where the stem sits.

11. Next, create the stem of the apple. Insert the knife into the center of the "island" in the second cake and cut out a small cylinder shape. The cylinder should be about 1 inch deep.

12. Add about 1 tablespoon of buttercream to the bowl with the crumbled cake. Add 3 to 4 drops of brown food coloring and combine thoroughly. Carefully press the brown cake mixture into the cylinder cavity that you created for the stem.

13. Break the red velvet cake into crumbs and place them into a large bowl. Add 2 to 4 tablespoons of buttercream and mix until fully combined to make a cake mixture (page 11). You're looking for a Play-Doh-like consistency.

14. Fill the spaces in both cakes with the red velvet cake mixture. The tops of both cakes should be even.

15. Spread a thin layer of white buttercream around the edge of the apple's bottom cake. Carefully turn over the apple's stem cake and lay it on top of the bottom cake, so that the apple (red cake) is joined at the center.

16. Cover the cake in a crumb coat (page 13) and chill for at least 1 hour in the fridge or 15 minutes in the freezer.

Frosting and Decorating

17. Tint the remaining buttercream red.

18. Remove the cake from the refrigerator. Cover the cake in a smooth layer of the red buttercream (page 12).

19. Place a #65 tip into a plastic disposable pastry bag, and fill the bag with the green buttercream. Then place a #4 tip into a plastic disposable pastry bag and fill it with the brown buttercream.

20. Pipe out some branches onto one side and the top of the cake. An easy way to do this is to trace out your design with a toothpick before you begin piping. If you make a mistake, simply smooth out the frosting with a small offset spatula.

21. Go in with green buttercream and randomly add leaves. Start by practicing leaves on a piece of parchment paper. Place tip on parchment and apply pressure to the bag. Slowly pull up and release pressure. Move to the cake and pipe a leaf onto a branch that you have already piped out.

make apple cupcakes!

Simply create little apples out of red velvet cake mixture. If you include stems, make them a little bigger than you'd think they should be. Prepare white cake batter and fill prepared cup-cake wells two-thirds full. Place a red velvet apple into each cupcake (some of apple will be visible) and bake as you normally would.

balloon cake

Not only does this cake have balloons on it, but it has a balloon in it. Not a real balloon, mind you, because that would be kind of gross, but a red velvet cake balloon. Now that, I *do* want to bite into!

My kids loved this cake and couldn't get enough of these miniballoons. I went to a local balloon shop and purchased six balloons and six holders for a grand total of $2.57. The kids were awestruck that they had their very own balloon stuck into their slice of cake! And I was awestruck that they were entertained for three hours and that I actually got to enjoy a piece of the delicious cake for once!

Okay, it was two pieces.

2 recipes White Cake, page 19

½ recipe Red Velvet Cake, page 21

1 recipe Basic Buttercream, page 24

Rainbow sprinkles

SPECIAL EQUIPMENT: *4 8-inch round cake pans; 9 × 13-inch pan; offset spatula; 4½-inch round cookie cutter; 2-inch round cookie cutter; small, sharp knife; 6 miniballoons and holders*

DIFFICULTY: *Medium*

Baking

1. Bake four white cake layers in 8-inch pans.

2. Bake the red velvet cake in a 9 × 13-inch pan.

3. Cool the cakes to room temperature.

4. Make the buttercream.

5. Set one white cake layer on a piece of parchment paper. Using the offset spatula, cover the top in a thin coat of buttercream, then add a second cake layer. Repeat with the other 2 cake layers to make a second 2-layer cake.

6. Refrigerate the cakes for at least 6 hours.

7. Prepare red cake mixture by crumbling the red velvet cake into a large bowl, removing any dark or hard spots as you crumble. Add 2 to 4 tablespoons of buttercream and combine to make a cake mixture (see page 11). You're looking for a Play-Doh consistency. Refrigerate the cake mixture in a plastic bag or storage container while you work on the white cakes.

Making the Surprise

8. Remove one cake from the refrigerator. Center a 4½-inch round cookie cutter on the cake and press in gently to make a guideline.

9. Center a 2-inch round cookie cutter on the guideline and tip it over to use as a scoop. Slowly pull the cutter around the top of the cake, keeping it centered on the guideline. Remove the excess cake and discard (or use it to make cake pops). It's important that the cake not become too crumbly or soft, so freeze it for 30 minutes or so if you're finding it hard to work with.

10. Repeat with the second layer cake.

11. Next you'll create the knot at the bottom of the balloon. Working with one of the two cakes, use a small, sharp knife to make a guideline through the bottom center of the channel you just created. Cut gently on either side of the guideline with your knife to make a minichannel. Use the tip of the knife to carve out the crumbs. To remove the crumbs, place your hand flat over the center of the cake and tip the cake over so the crumbs fall into the sink or garbage. This is much easier than trying to dig out each little crumb! This works best if your cake is very cold and solid.

12. Remove a golf ball–size portion of red velvet cake mixture and roll it into a snake about the size of the minichannel you just carved. Insert it

carefully into the channel, making sure it's level with the bottom of the larger channel.

13. Next, fill the entire cavity with red velvet cake mixture. This is the bottom of your cake— be sure to keep track. An easy way to keep track is to write "bottom" on the parchment or put a small dollop of buttercream on the outside of your cake.

14. Fill the other cake with cake mixture as well.

15. Place a thin layer of white buttercream around the top of the bottom layer (both the outside and the island), taking care not to get it on the red velvet cake mixture.

16. Carefully flip the top cake onto the palm of your hand and place it on top of the bottom cake layer. Try to align the two cakes as much as possible so that the red velvet mixtures meet evenly.

17. Cover the cake in a crumb coat (see page 13) and chill for at least 2 hours.

Frosting and Decorating

18. Cover the cake in a smooth coat of buttercream (see page 14). Assemble the balloons and holders and insert them into the cake.

19. Cover the top of the cake in rainbow sprinkles.

construction cake

2 recipes White Cake,
page 19

1½ recipes Basic
Buttercream, page 24

Orange, yellow, and black
gel food coloring

SPECIAL EQUIPMENT:
2 9 × 13-inch pans; cake leveler;
long, sharp knife; ruler;
offset spatula; disposable plastic
pastry bags; #3, #4, and
#7 decorating tips

DIFFICULTY: *Medium*

As a project manager for a construction company, my husband knows a great deal about construction. This is a fantastic quality when you happen to have three sons who are fascinated by machines. We'll drive by a big truck on the road and I'll say, "Look at the big truck, guys!"

My husband will respond, "That's a John Deere crawler mounted hydraulic boom backhoe excavator."

The kids will say, "Whoa. Cool, Dad!"

And I'll roll my eyes and mutter, "I totally knew that."

I totally did not know that.

Baking

1. Add a few drops of orange gel food coloring to one batch of white cake batter. Bake each cake in a 9 × 13-inch pan.

2. Freeze the cakes for at least 6 hours, preferably overnight. The cakes must be firm, not crumbly.

Making the Surprise

3. Level the cakes if they aren't perfectly flat (see page 9) and trim off the hard edges.

4. Prepare 1 recipe buttercream. Transfer 2 cups of buttercream to another bowl and tint it orange to match the cake. Leave the remaining buttercream white.

5. Cut both the orange and white cakes into 5 × 7-inch rectangles.

6. Using the offset spatula, cover one layer of each color with a very, very thin layer of orange buttercream for the orange layer and white buttercream for the white layer, cleaning the spatula between color changes.

7. Place the unfrosted orange layer on the frosted layer. Repeat with the white layers. Freeze the cakes if they're soft and crumbly.

8. Remove the orange cake from the refrigerator and place it on a sheet of parchment paper so that a long side faces you. Using a knife with a blade that's more than 5 inches long and starting at one end of the cake on the top corner, cut through the cake at a 45-degree angle.

9. Move 2 inches over and make another cut. Try to maintain the same angle throughout the whole cake. Repeat until you've reached the end of the cake. If the cake becomes too crumbly or soft, freeze it until it's firmer.

10. Make the same cuts on the white cake.

11. Place the lower corner of the orange cake on the right side of a cake stand or serving plate. Take the second piece of white cake and lean it against the orange cake. (You may frost between the layers if desired; the frosting should be very light and easy to spread.)

12. Add the next layer of orange cake, then the white cake, then the orange corner, until the cake is assembled.

13. Assemble a second cake with the remaining pieces. If you need only one cake for your party, wrap the extra one in plastic wrap and then in foil. It will keep in the freezer for up to 3 months.

14. Freeze the finished cake for at least 1 hour before frosting.

15. Cover the cake in a crumb coat or white buttercream (see page 13). Make sure your frosting is very pliable and easy to spread.

Frosting and Decorating

16. Cover the cake in a smooth coat of white buttercream (see page 14).

17. Insert a #4 tip into a pastry bag and fill it with the orange buttercream. Pipe out triangles resembling construction cones around the base of the cake.

18. Prepare a half-batch of buttercream. Transfer 1½ cups to another bowl and tint it yellow, tint the remaining buttercream black. Insert a #7 tip into a pastry bag, fill it with the yellow buttercream, and pipe out a large rectangle on your cake. Fill in the shape with the yellow buttercream, and use a small offset spatula to smooth out the frosting.

19. Use a toothpick to trace the sign you want to display onto the yellow buttercream. Then insert a #3 tip into a pastry bag, fill it with the black buttercream, and pipe out the design.

20. Chill until ready to serve.

21. Since the surprise-inside design is best seen on the long side of the cake, be sure to cut into that side for your first piece!

duck cake

2 recipes White Cake,
page 19

Yellow gel food coloring

1 recipe Basic Buttercream,
page 24

SPECIAL EQUIPMENT:
*2 8-inch round cake pans;
9 × 13-inch cake pan;
4½-inch round cookie cutter;
duck or similarly shaped cookie
cutter (I used a tugboat); small
offset spatula; tissue-paper pom-
poms (purchased or handmade)*

DIFFICULTY: *Medium*

Babies. Baby showers. Holding babies. Looking at babies. Thinking about babies. Snuggling with babies.

I sorta have a thing for babies. I've had five of them.

Babies and cakes have taken over quite a big section of my brain, so dreaming up this little gem was a given. The second I saw this cake I knew it would be perfect for a baby shower. And even though this particular cake is yellow, you could easily make it pink or blue for a gender-reveal party!

Baking

1. Bake one white cake recipe in two 8-inch layers.

2. Add yellow food coloring to the batter of the other white cake and bake it in a 9 × 13-inch pan. Cool all the cakes to room temperature.

Making the Surprise

3. Place the white cakes on a piece of parchment. Using a 4½-inch cookie cutter, find the center of one cake layer. Gently press in the cookie cutter to create a guideline. Repeat with the second white cake layer.

the guideline. Straighten it upright and slowly pull the cutter through the cake just as you did on the first layer, again being careful to maintain a uniform depth. Do not cut deeper than whatever reference you've chosen on your cutter.

8. Remove the excess cake and discard or save to make cake pops.

4. Center the cookie cutter on the guideline and insert it into the cake at an angle, then straighten it upright. Slowly pull the cutter through the cake, keeping it centered on the guideline the whole time and being careful to maintain a uniform depth. I used the boat deck portion of my tugboat cookie cutter as my reference, taking care never to push it deeper than that line.

5. Remove the excess cake and discard or save to make cake pops.

6. At this point make sure your cake is well chilled—not frozen, but not crumbly or too soft.

9. Since my cookie cutter is not shaped exactly like a duck head, I grabbed my favorite baby spoon and removed a bit more cake. In the picture I am smoothing out the back of the duck's head. The inner side is where the duck's bill would be, so focus on that when carving the bill.

10. Make the buttercream.

11. Trim off any hard or discolored spots from the yellow cake, then crumble it into a large bowl. Add about 3 tablespoons of buttercream and mix to create a pliable cake mixture.

7. Working on the second layer now, turn the cutter over and insert it so that it is centered on

12. Gently fill the spaces in the cake layers with pieces of cake mixture. Take care to press it up into the sides, especially into the narrow edges, like the duck's beak and tail. Make sure both cake cavities are completely full of yellow cake mixture to the height of the layer.

13. Using the offset spatula, place a thin layer of buttercream on the tops of the cake layers (white parts only). Place the bottom layer (the one with a wider band of cake mixture) on a cake stand. Gently turn the top layer over and center it on top of the bottom layer.

14. Cover the cake in a crumb coat (see page 13) and refrigerate it for at least 1 hour.

Frosting and Decorating

15. Tint the remaining buttercream a soft yellow to match the interior of the cake. Spread a thin layer over the cake and smooth it out with the offset spatula.

16. Decorate with tissue-paper pom-poms if you like. Check out the tutorial at hgtv.com/handmade/how-to-make-tissue-pom-poms/index.html.

house cake

Do you have a dream profession? Something that you would do simply because you love it and not for any financial reason? I do.

It's house flipping. I am addicted to every house-flipping show on TV, but have a special fondness for *Flip This House* and *Flipping Boston* on A&E and *Flip or Flop* on HGTV. Every Saturday morning I get up, get the kids breakfast, make a cup of coffee, and sit down in our squeaky rocking chair to feed the baby and turn on *Flip This House*. While it normally takes twenty minutes to feed my little angel, for some reason it takes at *least* an hour on Saturday mornings.

There's just something about taking an old, broken-down, and deteriorating home and making it new that thrills me. I like seeing things become new again.

Since I don't think I'll be buying an old home anytime soon, I will just stick to cake home construction for now!

Baking

1. Bake 4 layers of white cake in 8-inch cake pans. Cool to room temperature, then freeze them for at least 6 hours, preferably overnight.

2. Bake 1 layer of chocolate cake and 1 layer of red velvet cake in 8-inch round cake pans. Cool to room temperature.

3. Prepare the buttercream.

4. Crumble the red velvet cake into a large bowl, add 2 to 4 tablespoons of buttercream, and use a rubber spatula to combine them thoroughly. The cake mixture will have a smooth, Play-Doh-like consistency. In a separate bowl, repeat the process with the chocolate cake.

5. Place the chocolate and red velvet cake mixtures into airtight containers or plastic bags and refrigerate until you're ready to assemble the cake.

2 recipes White Cake, page 19

½ recipe Chocolate Cake, page 20

½ recipe Red Velvet Cake, page 21

1 recipe Basic Buttercream, page 24

Sky blue gel food coloring; white gel food coloring (optional)

SPECIAL EQUIPMENT:
4 8-inch round cake pans; rotating cake stand; offset spatula; bamboo skewer; 4½-inch and 4-inch round cookie cutters; small, sharp knife; disposable plastic pastry bag; #46 and #12 decorating tips; coupler

DIFFICULTY: *Medium*

Making the Surprise

6. Place a white cake layer on a piece of parchment paper. Using an offset spatula, cover the top smoothly with about ¼ cup of buttercream (see page 12). Top with another white cake layer. Repeat with the remaining 2 white cake layers to make a second 2-layer cake.

7. If you have a rotating cake stand or turntable, place one of the layer cakes on it. Now you'll be removing a cone-shaped piece of cake from the center of the cake, cutting three-quarters of the way through the cake. Insert a long, bamboo skewer into the center of the cake. Center the 4½-inch round cookie cutter on the cake, then gently press down, just far enough to make a guideline.

8. Insert a small, sharp knife on the guideline at about a 45-degree angle toward the center of the cake and press it into the cake until it hits the skewer. Slowly and steadily cut around the cake, keeping the base of the knife on the guideline and the tip of the knife touching the skewer.

9. When you've cut around the entire perimeter of the cake, gently remove the cone-shaped portion you just cut. Set aside the cone.

10. Place the other white cake on parchment paper. Center the 4-inch round cookie cutter on the cake and gently press down, just far enough to make a guideline. (You can use the 4½-inch cookie cutter; just move in ½ inch from the cutter to cut the guideline with your knife.)

11. Insert a thin knife on the guideline straight down to the bottom of the cake and follow the guideline to cut a perfect cylinder.

12. Remove the cylinder of cake. If you put the cone piece on top of the cylinder piece, you can start to see how the house will come together.

13. Crumble the cone and cylinder cake scraps into a large bowl, add 2 or so tablespoons of buttercream, and combine them thoroughly with

a rubber spatula to make a white cake mixture. Add white food coloring if you want a brighter white.

14. Remove the red velvet and chocolate cake mixtures from the refrigerator. Fill the triangular cavity (cone shape) of the first layer cake with the chocolate cake mixture, taking care to make a level top. Place this cake in the freezer.

15. Place the second layer cake onto a cake stand. Scoop out ¼ cup of the white cake mixture and carefully roll it into a rope that's about 1 inch in diameter. Cut off the ends to create a clean 2-inch-long cylinder.

16. Stand the small white cylinder on end in the center of the cavity and fill all around it with red velvet cake mixture. Cover the top of the small white cylinder with a ¼-inch-thick layer of red velvet cake mixture.

17. Now make a thin rope from the white cake mixture and cut it to a 6-inch length. Square off the edges with a knife so that it has 4 sides. Press one end against the other to make a circle. Center the circle on top of the red velvet cake mixture.

18. Gently fill the cavity with more red velvet cake mixture, taking care to make a level top.

19. Cover the top white part of this cake with a thin layer of white buttercream, taking care not to get any frosting on the red velvet cake mixture.

20. Remove the other layer from the freezer, then carefully flip it onto the palm of your hand and transfer it to the top of the layer on the cake stand. You've completed your house!

21. Cover the cake in a crumb coat (see page 13) and freeze it for at least 15 minutes, but preferably overnight.

Frosting and Decorating

22. Reserve 1 cup of the remaining buttercream to leave white (and feel free to tint it with white food coloring to make it brighter). Tint the rest sky blue and use it to frost the exterior of the cake (see page 12).

23. Insert a coupler into a pastry bag and attach a #46 decorating tip. Fill the pastry bag with white frosting. Pipe evenly spaced vertical lines around the bottom of the cake, making sure the serrated side of the tip is facing the cake, then finish the fence by piping white frosting in 2 rows across the vertical lines. Change to the #12 decorating tip and pipe out fluffy clouds.

24. Chill the cake until serving.

butterfly cake

2 recipes White Cake,
page 19

1 recipe Basic Buttercream,
page 24

Pink gel food coloring

Pink glaze, page 131, made
with a little less liquid for a
darker shade of pink

SPECIAL EQUIPMENT:
*2 6-inch round cake pans;
9 × 13-inch cake pan; 2-inch
round cookie cutter; small,
thin knife; offset spatula;
disposable plastic pastry bag;
#104 decorating tip*

DIFFICULTY: *Challenging*

I could totally fib here and tell you this cake is for my little girl. That she loves pink butterflies and adores the fantasy of dancing through fields with pink butterflies flitting behind her.

But I can't.

This cake is all for me. My favorite outfit when I was a little girl was a dress my mom sewed for me. It was a gray dress with a drop waist and a big, thick band that went across my hips. Sitting precariously on my hip was a pink butterfly. Oh, how special I felt wearing that dress!

These days I'm not sporting anything that accentuates my hips, but I can try to relive my fond memories in cake form.

Ironic, seeing as how the cake contributes to the hips.

Baking

1. Bake 1 white cake recipe in two 6-inch round cake pans (the layers may take a few extra minutes to bake).

2. Cool the layers to room temperature. Remove them from the pans and place each on a sheet of parchment paper. Stack the cakes and wrap them in plastic wrap. Freeze for no less than 6 hours, preferably overnight.

3. Remove the cakes from the freezer, but leave them covered in plastic wrap until ready to use. Bake the second white cake recipe in a 9 × 13-inch cake pan.

4. Level the cakes if needed (see page 9).

Making the Surprise

5. Make the buttercream.

6. Make the pink glaze.

7. Center a 2-inch cookie cutter on top of one layer and gently press down to create a guideline. Repeat on the second round layer.

8. Insert a small, thin knife about 1 inch into the center of your cake at a 45-degree angle. Cut around the cake as far as the guideline. Remove the excess cake and crumbs.

9. Using a small spoon or dull knife, start carving around under the guideline, pushing the spoon slightly under the line and keeping a peak in the middle. This is the "small wing," or lower portion, of the butterfly and thus the bottom of the cake.

10. Repeat these cuts on the other round layer, only make the cavity wider and slightly deeper. This is the "large wing," or top, layer.

11. Make a cake mixture using the 9 × 13-inch cake (see page 11). Be sure to remove any dark or hard spots prior to crumbling. Crumble three-quarters of the cake into a large bowl and mix in 3 to 6 tablespoons of buttercream and 2 to 4 drops of pink gel food coloring. To add eyes to the large wings, crumble the remaining cake into a medium bowl and mix in 1 to 2 tablespoons of buttercream.

12. To get the dark rim you see on the lower wings, pour ¼ cup of the pink glaze into the well of the bottom layer of the cake. Set the cake aside so the glaze can settle evenly.

15. Roll a small piece of the white cake mixture into a snake about 1 inch thick.

13. Roll a small piece of the pink cake mixture into a snake. Softly flatten the snake in your hand and gently press it into the sides of the "large wing" (top) layer and up the center point. Repeat with more pink cake mixture so that the hole and point are covered in a layer of pink.

14. Repeat with the second layer, taking care not to disturb the pink glaze.

17. Place a thin layer of white buttercream around the top outside rim of the "small wing" (bottom) layer, taking care not to get any frosting on the pink part, then transfer the bottom layer onto a cake stand. Gently turn over the "large wing" (top) layer and place it on top.

18. Freeze the cake for at least 1 hour, then cover it in a crumb coat (see page 13).

Frosting and Decorating

19. Tint the remaining buttercream pink (except for a tiny bit for the butterfly) and cover the cake in a smooth coat of pink buttercream (see page 14). Using a #104 tip, pipe out a white butterfly (4 wings) to sit on the edge of the cake. Or add more butterflies for more impact!

16. Tuck the snake into the "large wing" (top) cake layer. Then go back and fill in the open areas of both layers with pink cake mixture.

20. Chill the cake until serving.

daisy cake

This sweet little cake design was originally inspired by my mother-in-law. Her birthday is January 6, and she loves flowers. Since her birthday is in the dead of winter, I decided to make her a cake with glorious happy flowers all over it! The phrase hidden in the grass in this cake is "Oh Happy Day!"

The beauty of this cake is that it is so accommodating. You can write any phrase that is significant to you; you can make the flower inside any color or flavor. Just think of how you could bless someone with a truly personalized and delicious cake!

I found a flower cookie cutter to use as a visual guide of what I wanted to create. Sometimes props can help us visualize!

1 recipe Chocolate Cake, page 20

½ recipe White Cake, page 19

1 recipe Basic Buttercream, page 24

½ recipe Chocolate Buttercream, page 25

White, yellow, sky blue, and green gel food coloring

SPECIAL EQUIPMENT: *2 6-inch round cake pans; 9 x 13-inch cake pan; 3-inch round cookie cutter; small, sharp knife; small offset spatula; disposable plastic pastry bag; #2, #3, and #4 decorating tips*

DIFFICULTY: *Challenging*

Baking

1. Bake the chocolate cake in two 6-inch round cake pans. Cool to room temperature, then freeze for no less than 6 hours, preferably overnight. (You may need to extend the recommended cooking time when using 6-inch pans.)

2. Bake the white cake in a 9 × 13-inch pan. Cool to room temperature.

Making the Surprise

3. Make the white buttercream.

4. Level the chocolate cakes.

5. Center the 3-inch round cookie cutter on top of one cake layer and gently press down to make a guideline. Repeat with the other cake layer.

7. Next, carve out the bottom petal. Using a small, sharp knife, cut out a small circle in the island at the center of the cavity. Remove all the crumbs.

6. Using a small spoon (I got out my favorite silver baby spoon), insert it at a 45-degree angle and cut 1 inch deep just inside (but not cutting into) the guideline. Work your way around the cake, carving out small sections at a time until you have an even well with an island in the center. Clear out all the crumbs and extra cake.

8. Repeat the carving on the second chocolate layer so that the two layers match.

9. Use about ¼ of the white cake to make a white cake mixture (see page 11). Break the cake into crumbs in a large bowl and mix in 1 scant tablespoon of buttercream until the mixture is the consistency of Play-Doh. Add 1 or 2 drops of white food coloring if needed.

10. To fill in the cavities in the chocolate layers, roll out 2 small balls of white cake mixture for the centers and "snakes" as needed to fill the outside rims.

11. Gently press a ball into the center of the first cake layer and snakes into the outside rim.

12. Crumble ¼ cup of the white cake into a bowl, mix in 1 small teaspoon of buttercream,

and add 1 or 2 drops of yellow food coloring. Roll this cake mixture into a ball and place the ball in the center of the white cake mixture. Build up white cake mixture around the ball.

13. Fill the other chocolate layer with white cake mixture. Create a small well where the top of the yellow ball will fit. Go back in and fill any open areas with the white cake mixture.

14. Make the chocolate buttercream. Using the offset spatula, spread a thin layer around the top of the chocolate part of one layer, being careful not to get any frosting on the white or yellow cake mixture. Gently turn over the other cake and lay it on top of the first so that the white petals meet in the middle.

15. Set the cake on a cake stand and cover it in a crumb coat (see page 13). Chill for at least 1 hour.

Frosting and Decorating

16. Divide the white buttercream between 2 bowls. Tint one white and the other sky blue. Set aside ¼ cup of the white buttercream and tint it yellow. Set aside another ¼ cup and leave it white. Tint the remaining buttercream green.

17. Cover the exterior of the cake in a smooth layer of skyblue buttercream (see page 12).

18. Insert a #2 decorating tip into a pastry bag and fill it with the green buttercream. Go around the base of the cake and pipe out the words "Oh Happy Day!" (or whatever phrase you like!) over and over again. To create the effect of grass, simply pipe stacks of the same letter. Above the *A* write four more *A*s; above the *Y* write two more *Y*s; and so on. Go around the entire cake and vary the heights of the letter stacks.

19. Insert a #3 tip into a pastry bag and fill it with white buttercream. Pipe out six-petal daisies all around the sides of the cake. If you created a smooth top, you can put them on top as well. I left the top of my cake rough.

20. Insert a #4 tip into a pastry bag and fill it with the yellow buttercream. Place a dot in the center of each flower.

21. Chill the cake until serving.

paw print cake

I am a dog person. I started begging my parents to get me a dog when I was about eight. When I was thirteen, my parents heard about a dog that was tied to a leash and stuck outside all day while the owners were at work. Apparently the neighbor kids kicked and tormented the dog, and when the owners found out, they decided to give him away so he wouldn't be abused anymore. Two-year-old Nicholas, a Lhasa Apso–terrier mix, made his way into our lives and hearts.

He was a much-loved and pampered pet and lived a wonderful life up until he passed away a few years ago. I fondly remember him as the first dog to touch all our lives and firmly define us as dog people.

We now have a Yorki-Poo mix named Dodger. Little Dodger, the sweet and gentle dog who lets babies tug his ears and chases after mischievous little boys. I love seeing Dodger interact with my family, and I hope my kids will grow up to be dog people, too.

Just to be clear, this cake can totally work as a cat paw . . . so it's perfect for cat people, too.

Baking

1. Bake the white cake into four 8-inch layers. Cool to room temperature, then freeze for at least 6 hours, preferably overnight. Level all the layers (see page 9).

2. Bake the chocolate cake in a 9 × 13-inch pan and cool it to room temperature.

3. Prepare the buttercream.

4. Remove any dark or hard edges from the chocolate cake, then crumble it into a large bowl. Mix in 2 to 4 tablespoons of the buttercream to create a cake mixture (see page 11). You're looking for a Play-Doh-like consistency. Cover the bowl with plastic until ready to use.

---

2 recipes White Cake, page 19

1 recipe Chocolate Cake, page 20

1 recipe Basic Buttercream, page 24

Brown gel food coloring

SPECIAL EQUIPMENT: *4 8-inch round cake pans; 9 × 13-inch cake pan; 2-inch round cookie cutter; small offset spatula; 3-inch round cookie cutter; disposable plastic pastry bags; coupler set; #4 decorating tip and grass tip*

DIFFICULTY: *Challenging*

Making the Surprise

5. Place one white layer on a cake stand and a second white layer on a piece of parchment. Center a 2-inch round cookie cutter on the cake and press lightly to create a guideline. Repeat on the second white layer.

6. Using a small spoon, hollow out a shallow well (no more than 1 inch deep) in the center of the layer on the cake stand, leaving a little island of cake slightly raised in the center.

7. Now turn to the second white layer. Hollow out a well that's a little deeper than the previous hole, and this time, don't make an island.

8. Fill both cavities with chocolate cake mixture.

9. Using the offset spatula, place a thin layer of buttercream around the edge of the bottom layer, avoiding the cake mixture. Gently turn over the second layer and place it on the bottom layer so that the cake mixtures meet in the middle. This is the bottom of your paw print. If the cake seems too soft or crumbly, freeze it for at least 30 minutes.

11. Make the same guideline on one of the remaining layers.

10. Center a 3-inch round cookie cutter on top of the 2-layer cake. Press in slightly to create a guideline.

tips and tricks

To keep your channel even, place a piece of tape around the spoon to show the desired depth. Simply insert the spoon into the cake until you reach the tape and dig out the cake.

12. Working on that layer, center a spoon on the guideline and create a shallow channel in the cake.

13. Cut a matching channel into the top of the 2-layer cake. Fill the channel on top of the 2-layer cake with cake mixture (mound it a bit, since you've carved out cake in the other layer and it needs to fill the space).

14. Using the offset spatula, spread buttercream around the top of the edges and center of the 2-layer cake.

15. Turn over the white layer with the channel, and place it on top of the 2-layer cake, settling it gently so that the cake mixture fills the channel.

16. Using the 2-inch cookie cutter, repeat steps 13 to 15 with the three-layer cake on the cake stand and the remaining layer of white cake.

17. Cover the cake in a crumb coat (see page 13). (If the cake is too soft or crumbly, freeze it first.)

Frosting and Decorating

18. Tint the remaining buttercream light brown. Set aside ½ cup. Using a small offset spatula, cover the cake in a smooth coat of buttercream (see page 14).

19. Add a little more brown food coloring to the brown buttercream that was set aside.

20. Insert a coupler into a pastry bag, then attach the #4 decorating tip. Fill the bag with the dark brown buttercream. Make small blob outlines all over the cake. Remove the #4 tip and attach the grass tip. With small, short motions, add a fur-like texture to the blobs. You might want to practice this motion on parchment first.

21. Chill the cake until serving.

CELEBRATING YOU

i *have* this theory about bakers. It seems to me that they're bountiful givers who get just as much joy from giving out their baked goods as they do from creating them, if not more.

I love that about all you amazing bakers and wanted to show you how much I appreciate it by helping you create cakes that pack a visual punch.

neapolitan hi-hat cake

1 recipe Decadent Brownies, page 23

1 recipe Strawberry Cake, page 22

½ recipe Chocolate Buttercream, page 24

2 16-ounce containers of whipped topping, such as Cool Whip, very well chilled

12-ounce bag of milk chocolate chips

½ cup canola oil

SPECIAL EQUIPMENT: *9-inch round cake pans; cake leveler or long serrated knife; disposable pastry bag; small offset spatula*

DIFFICULTY: *Easy*

Every day I teach my kids not to be selfish:

"Put others' needs first."

"Give one to your sister before you eat them all."

"Share with him the way you want him to share with you."

On the day I made this cake, I cast selflessness to the wind. I couldn't help it. I love brownies. I love strawberries. I love mounds of sugary whipped frosting. I love milk chocolate.

I love this cake. And I wanted it all for myself.

It's probably the easiest cake in the book. If you can turn on a mixer, you can do it. Heck, you don't even need a mixer. Just give it a shot. I promise you won't be disappointed.

Baking

1. Bake the brownies in a 9-inch round cake pan.

2. Bake the whole strawberry cake recipe in a single 9-inch pan. You may need to add extra time for baking.

3. Cool both the cake and the brownies to room temperature.

4. Prepare the chocolate buttercream.

5. Level the strawberry cake to the height of the brownies (see page 9). A hand-held leveler works best, but you can also use a long serrated knife.

Making the Surprise

6. Place the brownie layer on a cake stand, then set the strawberry layer on top.

7. Using the offset spatula, cover the sides of the cake with chocolate buttercream (see page 13).

8. Place the whipped topping in a pastry bag and cut off the tip so that you have an opening about 1 inch wide. Pipe out a circle of whipped topping around the perimeter of the cake. This wall will help to keep the whipped topping in place.

9. Scoop extra whipped topping onto the cake and smooth it with an offset spatula.

10. Moving slightly in, pipe out another circle on top of the layer of whipped topping. Fill the center with more whipped topping. Continue this process until you have a nice point on top. If the whipped topping is not holding the layers, remove what you have and chill it until it's sturdier. You can then start again. Smooth the sides, if necessary, when you are done.

11. Freeze the cake for 30 minutes before coating it in chocolate; if you won't be serving the cake quickly, keep it in the fridge overnight.

Frosting and Decorating

12. Pour the chocolate chips and the canola into a heat-safe bowl. Heat 1 cup of water in a small saucepan over medium-low heat (you want a simmer, not a rolling boil). Set the bowl of chocolate over the saucepan and heat, stirring, until the chips are melted.

13. Pour the chocolate into a high-sided container. Let it sit for a few minutes to cool down.

14. Lay 4 pieces of parchment around the base of the cake. Pour the chocolate over the cake, coating the whole thing.

15. Let the chocolate set for about 5 minutes, then remove the parchment paper. (You can remove the paper earlier if you like the look of chocolate puddles—I kind of do!)

16. Keep the cake chilled in the fridge until ready to serve. Dip a knife in hot water before using it to cut the cake.

opposites cake

This cake is perfect for me because it pairs rich, decadent chocolate cake with a light and moist vanilla cake. It's my go-to cake when I can't decide which craving I want to fulfill.

This is also a perfect man-pleasing cake; the lack of frill and fuss makes it an excellent choice for anyone who prefers simplicity and flavor to over-the-top design. *Not that there's anything wrong with over-the-top design.*

SPECIAL EQUIPMENT:
2 9-inch round cake pans; offset spatula

DIFFICULTY: *Easy*

Baking

1. Bake the whole white cake recipe in a single 9-inch round cake pan and the whole chocolate cake recipe in another 9-inch pan (the cakes may take a few extra minutes to bake). Cool the layers to room temperature.

2. Prepare the buttercreams.

3. You'll be pairing vanilla cake with chocolate frosting and vice versa, so think about which flavor of frosting you want on top. If it's vanilla, use the white cake as the bottom layer, and vice versa. (I choose white cake on the bottom and vanilla buttercream on the top, but this tough decision is in your hands!)

Making the Surprise and Frosting

4. Place the white cake on a cake stand and cover the top and sides with chocolate buttercream (see page 12). Then freeze the cake for at least 30 minutes and up to 2 hours.

5. Cut four 2 × 6-inch strips of parchment paper. Remove the cake from the freezer and gently lay the parchment strips around the edges of the cake, covering them completely.

6. Place the chocolate layer on top and cover it in vanilla buttercream.

7. Gently remove the parchment strips. You should have clean edges between the two layers!

8. Serve the cake immediately.

oreo cake

This cake makes me happy in every way. Not only is it easy to bake, but it's fun, looks cool, and is insanely delicious! I recommend making it for a large group of people so that you're not tempted to eat more than one piece. 'Cause you (and by you I mean me) *will* be tempted. And then you'll take a picture of it and post it on Instagram and ask your friends if you should eat just one more piece. And you *know* that you have the smartest, most encouraging friends, so when they say, "Yes! Eat it!" you'll wholeheartedly agree. And then you'll cut just one more piece and ask your friends on Twitter. And then just one more piece and ask your friends on Facebook.

And that is how to get fat from social media.

#justsayin

oreo filling

¼ cup (½ stick) unsalted butter

¼ cup shortening

3 cups powdered sugar

2 to 3 tablespoons milk

2 teaspoons vanilla extract

Place the butter and shortening in a mixer and, using the paddle attachment, mix until fully combined. Add the sugar, 1 cup at a time, alternating with the milk and vanilla. The filling should be very thick, but you can add more milk if you need to.

Baking

1. Lay 19 Oreos in an even layer on the bottom of a 9-inch round cake pan. Repeat in a second 9-inch pan.

---

1 recipe Chocolate Cake, page 20

1 15.25-ounce package Oreos

1 recipe Oreo Filling

1 recipe Chocolate Buttercream, page 25

1 premade Oreo piecrust

SPECIAL EQUIPMENT: *2 9-inch round cake pans; offset spatula; fondant smoother (optional)*

NOTE: *If you can't find a premade Oreo piecrust, you can make your own Oreo crumbs. Just remove the filling from 24 Oreos and pulse the chocolate cookies in a food processor until they're ground fine.*

DIFFICULTY: *Easy*

2. Prepare the chocolate cake.

3. Divide the batter evenly between the cake pans, covering the Oreos thoroughly. Bake as you normally would.

4. Cool the layers completely.

5. Make the Oreo filling and chocolate buttercream.

Making the Surprise

6. Place 1 cooled layer on a cake stand, Oreo cookie side down. Using an offset spatula, cover the top with a 1-inch-thick layer of Oreo filling.

7. Top with the other layer, Oreo cookie side up.

Frosting and Decorating

8. Cover the cake in Chocolate Buttercream.

9. Working quickly, cover the cake in Oreo crumbs. Pour a small amount of crumbs on top of

the cake near the edge. Use one hand to move the crumbs over the edge and the other to catch the crumbs and press them into the side of the cake. Repeat to go around the whole cake. When done with the sides, pour the excess crumbs back onto the top of the cake. Make sure you have a smooth layer on the top and sides. I pressed a fondant smoother into the cake to create clean edges, but if you do not have this tool you can use your hand.

zebra cake

This cake is dedicated to Sunny Mabrey. I first discovered her through her six-second videos on the smartphone app called Vine, but she is also an accomplished actress on the big and small screen. I have to tell you, she is funny. Like see-her-Vine-on-Monday-then-laugh-out-loud-on-Thursday-just-thinking-about-it funny.

So why a zebra cake for Sunny? I think she is a closet zebra print fan. I also think she would appreciate something that appears classic and harmless, but that secretly disguises a wild and crazy uniqueness, just like my chocolate-covered zebra cake!

This cake would also be perfect for birthdays, for bachelorette parties, and for anyone who has a wild side!

SPECIAL EQUIPMENT: *4 8-inch round cake pans; serrated knife; small offset spatula*

DIFFICULTY: *Easy*

Baking

1. Bake 2 layers of white cake and 2 layers of chocolate cake in 8-inch round cake pans. (You'll only need one layer of chocolate for this cake. Cover the remaining layer in plastic wrap, then foil, and freeze for up to 3 months—unless you want to eat it now!)

2. Cool the layers to room temperature, then freeze the 3 that you're using for 1 hour or refrigerate them for several hours.

Making the Surprise

3. Place all 3 layers on parchment paper.

4. Turn over one of the white layers. Using a sharp knife, cut out a cone shape from the center of the overturned cake. Your knife should go all the way through the layer of cake and the base of the cone should be at least 2 inches across.

5. Center the cone point side up on a cake stand.

6. Moving back to the layer of white cake you just cut the cone from, flip the layer over so the right side is facing up.

8. Using a serrated knife, start trimming layers of chocolate cake off, much as you would trim a turkey. Place small slices of chocolate on top of the white cake cone on the cake stand, covering it but not in a particular pattern.

7. Gently carve around the exterior of that same layer, leaving a curved cap of cake (bottom left of photo). Remove the top and set aside.

9. Now place the center section of the cut white layer (the "cap") on top of the chocolate layer.

10. Now trim thin layers of chocolate cake and place them on top of the white cake, covering the cake.

11. Place the top of the cut white layer over the chocolate cake layer.

12. Repeat this process with the second white layer. You won't need to create the small center cone in the second white layer. As you build the layers, keep in mind that you want the cake to be slightly raised in the center.

13. Your bare-bones cake will look pretty haphazard. That's okay! I was nervous the first time I made this cake, too.

14. Place the cake in the freezer for a couple of hours, then cover it in a crumb coat of chocolate buttercream (see page 13).

Frosting and Decorating

15. Cover the cake with chocolate buttercream, then use a small offset spatula to make random swoops in the buttercream. The surface can be as rough as you want.

leopard cake

This cake is inspired by the Real Housewives—specifically the Real Housewives of New Jersey. At first I was just fascinated by how they talk, their fashion sense, and their way of life. But lately I've become a bit more invested! When Melissa and Kathy joined the show, I truly became wrapped up in the well-being of the Jersey girls. I just want everyone to get along! To make skinny Italian meals and drink blk. water and attend lavish parties with over-the-top and fabulous dessert bars!

I know, I know. What fun would the show be if everyone got along? But I don't know if I can take any more fighting. My nerves are shot!

This is my little contribution to the happiness of the Real Housewives of New Jersey. If I ever had the opportunity to visit them when they're on a media tour, I'd bring them this cake.

Everyone feels better with cake, right?

2 recipes White Cake, page 19

Brown gel food coloring

1½ recipes Basic Buttercream, page 24

Edible gold spray (found at Walmart and most grocery stores)

SPECIAL EQUIPMENT: *2 8-inch round cake pans; plastic disposable pastry bags; offset spatula; #70 or #104 decorating tip*

DIFFICULTY: *Medium*

Baking the Surprise

1. Bake one white cake recipe in two 8-inch pans and set aside.

2. Prepare the batter for the other white cake.

3. Place 1 cup of batter in a bowl and mix in 1 or 2 drops of brown food coloring. Place another 1 cup of batter in a separate bowl and mix in 5 or 6 drops of brown food coloring. (You can add more or less food coloring—you're looking for dark and light brown batters.)

4. Place all three colors of batter in disposable plastic pastry bags and secure.

5. Begin making the leopard layer by preparing an 8-inch round cake pan. Trim a piece of parchment to fit snugly in the bottom, then spray the entire inside of the pan with baking spray.

6. Cut off a small tip of the dark brown bag and make a circle around the bottom of the pan.

7. Cutting off a larger tip from the white cake batter bag, carefully cover the brown circle with white batter.

8. Now go back in with the dark brown batter and make two circles—one in the center and one closer to the edges. Cut off a small tip from the light brown batter and pipe over those circles.

9. Cover with another thin layer of white batter.

10. For the final layer of dark brown batter, I make one circle toward the middle of the pan and one toward the outer edge. Pipe light brown batter over the top and cover with the remaining white batter.

11. You can also go back and use all remaining brown batters now. Just make random circles around the cake.

12. Bake this layer for 25 to 35 minutes at 350°F. Since this is a lot of batter for an 8-inch cake, be sure to check it at the 20-minute mark. If the edges are really browning but the center is still very loose, place a ring of foil around the edge of your cake. (Do your checking very quickly or with the oven door closed!)

13. Cool the layer completely.

14. Prepare the buttercream.

15. When you're ready to assemble the cake, place 1 plain white layer on a cake stand. Cover it in about 1 cup of buttercream.

16. Set the leopard layer on top and cover with about 1 cup of buttercream.

17. Top with the other plain white layer and cover the cake in a crumb coat (see page 13).

Frosting and Decorating

18. Insert a #70 or #104 decorating tip into a pastry bag and then fill it with buttercream. I used a #70 Ruffle Tip from BakeryCrafts. You can also achieve this look with a #104 tip from Wilton.

19. Place the tip at the bottom of the cake, and touching it gently to the cake, apply pressure to the pastry bag and move it up. Wrap the ruffle over the top of the cake by at least 1 inch. For the next row, start at the bottom again just to the right of the first ruffle. Make sure to overlap so that just the ruffles show. For the top of the cake, simply start at the outer edge, making sure you overlap so just the ruffle shows, and move around the cake until you reach the center.

20. Place the cake stand on wide sheets of newspaper or cut-up grocery bags (or do this outside). Spray the cake with the edible gold spray from 4 or 5 inches away, applying as much or as little as you want for your desired effect.

21. Chill the cake until serving.

cowboy boot cake

2 recipes White Cake,
page 19

1 recipe Chocolate Cake,
page 20

1 recipe Basic Buttercream,
page 24

Royal blue and black
gel food coloring

SPECIAL EQUIPMENT:
*4 8-inch round cake pans;
9 × 13-inch pan; large and
small offset spatulas; 2-inch,
3-inch, and 4½-inch round
cookie cutters; small, sharp knife;
disposable pastry bag;
#4 decorating tip*

DIFFICULTY: *Challenging*

This cake is dedicated to one of my very favorite people, the Pioneer Woman, blogger Ree Drummond, who's been a blessing to me in many ways. Her help and advice throughout the process of writing this book has been invaluable. Through Ree's blog I learned how to hold my camera, how to make jaw-dropping food, and that it's okay to be in love with your dog! The surprise is cowboy boots: The outside is meant to look like denim (if you know Ree's site, you'll understand the significance there!), and the cut piece is sitting in a little cast-iron skillet, one of her favorite kitchen accessories!

This is a pretty tricky cake to make, but I think it's fun to honor someone special in your life by personalizing a cake to fit *their* favorite things!

Baking

1. Bake 4 layers of white cake in 8-inch round cake pans.

2. Prepare the buttercream.

3. Cool the layers to room temperature and place one of the layers on a sheet of parchment paper. Level the cake if needed (see page 9).

4. Using an offset spatula, cover the cake layer with a thin layer of buttercream. Place another cake layer directly on top and freeze for at least 6 hours, preferably overnight. You're looking for a firm, noncrumbling cake.

5. Repeat with the remaining 2 white layers and freeze.

6. Bake the chocolate cake in a 9 × 13-inch pan, then cool it to room temperature.

7. Remove any hard edges, then crumble the chocolate cake into a large bowl. Add 1 to 3 tablespoons buttercream and mix to create a cake mixture (see page 11). You're looking for a Play-Doh-like consistency.

Making the Surprise

8. Working with one of the 2-layer cakes, center the 2-inch cookie cutter and press lightly to make a guideline. Repeat with the 3-inch and 4½-inch cutters.

9. Insert a sharp, thin knife into the 2-inch guideline and cut around the cake, following the guideline, until you reach buttercream (you cut through the whole first layer). Next, insert the knife at the 3-inch guideline and do the same thing, cutting through the first layer. Carefully go in and remove all the cake between the cuts, to give you a deep, round channel.

10. Make sure both cakes are well chilled at this point. If they seem at all crumbly, return them to the freezer.

11. To protect it from damage, you may remove the remaining center section of the cake before you do any more carving. It should easily separate from the bottom layer of cake. It does need to be replaced when done, so if you choose to remove it while carving out the next layer, be gentle and set it somewhere safe.

12. When you look at the profile of a cowboy boot, there's a space between the heel of the boot and the front sole. You'll work on this area next. Remove 1 cake from the freezer and set it on parchment. Insert your knife at the 3-inch guideline at a 45-degree angle toward the outside of the cake, then cut around the entire cake at this angle. You will cut through a single layer.

13. Remove the knife and insert it in the 4½-inch guideline at a 45-degree angle toward the outside of the cake. Cut around the entire cake, working as slowly as you need to.

14. Using a baby spoon or a butter knife, carefully remove the cake between the two cuts. This is what it looks like when you first go in and start removing excess cake. You want a rounded toe to your boot.

15. Remove as many crumbs as you can, then place the cake back in the freezer.

16. Remove the second 2-layer cake from the freezer. For this cake we are forming the tall part of the boot that goes over the ankle and calf.

17. Find the center of the cake using a 2-inch round cookie cutter. Press in gently to make a guideline. Repeat with a 4½-inch round cookie cutter.

18. Insert a thin, sharp knife into the 2-inch guideline and cut through one layer of cake, working around the guideline and keeping the knife straight up and down. Repeat the cut on the 4½-inch guideline.

19. To make the notch at the top of the bootleg, start at a center point between the two guidelines, insert the knife at a 45-degree angle, and cut to the edge of the inner (2-inch) guideline. Cut around the entire cake, going through the top layer only.

20. Repeat, with the knife angled in the opposite direction.

21. Remove the cake in this section except for the upside-down V that you carved in steps 19 and 20. Then gently fill the cavity with chocolate cake mixture.

22. Return to the other 2-layer cake (the lower part of the boot), and fill the cavity with chocolate cake mixture, starting with the heel of the boot and carefully building from there.

23. Place this cake on a cake stand and cover the edge and center of the white cake with a thin layer of buttercream, avoiding the chocolate cake mixture. Turn over the other cake and set it carefully on top. If the cake is crumbly, freeze it for at least 1 hour.

24. Cover the cake in a crumb coat (see page 13).

Frosting and Decorating

25. Tint the remaining frosting dark denim blue. I used about 10 drops of royal blue gel food coloring, then added 1 small drop of black.

26. Using an offset spatula, cover the cake in a smooth coat of denim blue (see page 14). Tint 1 cup of frosting a darker blue, with about 3 drops more food coloring.

27. Insert a #4 decorating tip into a pastry bag and fill it with the dark blue frosting. Pipe a thread detail along the edge of the top and the upper and lower edges of the cake. Then draw out a pocket shape on the side of the cake.

28. Using a small offset spatula or a butter knife, fill in the pocket outline with about 2 tablespoons of the darker frosting. Go back over with the #4 tip and pipe the thread detail.

29. Chill the cake until serving.

herringbone cake

2 recipes White Cake,
page 19

1½ recipes Basic
Buttercream, page 24

Black and yellow gel
food coloring

SPECIAL EQUIPMENT:
*2 11 × 16-inch sheet pans; long,
serrated knife; offset spatula;
ruler; disposable plastic pastry
bags; #104 decorating tip*

DIFFICULTY: *Challenging*

I'm totally smitten with the yellow and gray color concept of this cake. It's so classic and versatile, lending itself to any season or style. Add in the simple complexity (yeah, I said simple complexity) of the herringbone pattern and this cake shoots right up to the top of my favorites list.

I have to admit that this cake is totally inspired by the large flower boutonniere trend that was popular years ago. But that's me ... typically a few years behind trend!

Special note: *I'd consider this one of the more challenging cakes in the book due to the level of math and precision involved. I highly recommend reading through all the instructions prior to starting. You'll want to make sure you have all the tools and equipment necessary to ensure a fun and creative experience. Rest assured, I've made this cake a number of times and it gets easier every time. It's a great cake to make once, get the feel for, then experiment with. Use your favorite colors, different flavors ... create a rainbow design! But most of all, have fun with it!*

Baking

1. Bake 1 white cake in a well-prepared 11 × 16-inch sheet pan (see page 11).

2. Prepare the batter for the other white cake and mix in 1 or 2 drops of black gel food coloring. Bake a gray cake in the same size sheet pan. Let both cakes cool completely.

Making the Surprise

3. Carefully remove the gray cake from the pan and place it on a large piece of parchment. Using a long serrated knife, cut off the edges so that you have a perfect rectangle.

5. Cut the 10-inch square into ten 2 × 5-inch sections.

6. Trim the 6-inch piece so that it's 5 × 10 inches.

4. Cut one 10-inch square out of the sheet cake. You'll have a 6-inch piece remaining.

7. Cut the 5 × 10-inch section in half and place the layers on top of each other. You should now have a double-thickness 5 × 5-inch piece.

8. Remove the white cake from its pan and repeat the cuts.

9. On the 5 × 5-inch section, measure in 1 inch and place the knife on the cake at a 45-degree angle. Cut down, facing out, to form a triangle.

10. Now face the knife in the other direction, and cut down at a 45-degree angle.

11. You have now made two cuts that create a single three-dimensional triangle from the cake. Moving down the cake 1 inch, make the cuts again to give you two of these sections.

12. Repeat the cuts on the 5 × 5-inch gray cake. Pop your cake back into the freezer if it starts to get crumbly. This process works best if your cake is very chilled.

13. Prepare the buttercream. You will want a very soft and spreadable consistency of buttercream. To achieve this, add more liquid than the recipe calls for, 1 tablespoon at a time.

14. Place 2 white triangle sections on a cake stand about an inch apart. Cut a white triangle down the middle and use it to fill in the ends.

15. Lay 2 small gray rectangles flat on the right sides of the 2 white triangles.

16. Next, lay 2 small gray rectangles on the left side of the triangles to make a classic herringbone pattern. This is the bottom layer of your cake.

17. Using an offset spatula, place a thin layer of buttercream on the seams of the gray pieces to act as glue.

18. Lay 4 pieces of white cake in the same pattern on top of the gray layer, and seal the seams with a thin layer of buttercream. Repeat, alternating gray and white layers of cake, until the cake reaches your desired height.

19. Lay a triangle into the center of the cake (use the opposite color of the top layer). If it doesn't fill the space for any reason, simply fill in with scraps.

20. Cut another triangle piece down the middle and invert the 2 pieces to create the edges of the top layer.

21. Next, use a dab of frosting or a piece of tape to mark the front of the cake so you know where to place your buttercream flower. Aside from being beautiful, the flower will remind you where to make your first cut for serving, since the herringbone design is best seen when you cut crosswise into the pattern.

22. Freeze the cake for at least 6 hours.

23. Fill in the open notches, creating even sides, with room-temperature buttercream. Smooth out the buttercream to create a crumb coat (see page 13).

24. Freeze the cake until it has set.

Frosting and Decorating

25. Tint the remaining buttercream yellow and, using an offset spatula, cover the cake in a smooth coat (see page 14).

26. Insert a #104 decorating tip into a pastry bag, then fill it with the yellow buttercream. To make the flower at the front of the cake, pipe out petals (see the method for Ombre Cake, page 37), starting with the outer circle and moving in.

27. Chill the cake until serving. Cut the cake as shown in the first photo on page 120 to reveal the pattern.

CELEBRATING LOVE

i am a sucker for love. Not only do I love to feel love, but I love seeing it, anticipating it, and reveling in it. These cakes are sure to help you share the love with your loved ones!

candle rose cake

SPECIAL EQUIPMENT:
*empty aluminum cans of
varying sizes; 8-inch round
cake pan; serrated knife; pan
wire rack; baking sheet; offset
spatula; 9-inch round cake pan;
disposable pastry bags; 1M and
20 star decorating tips; coupler;
birthday candles*

DIFFICULTY: *Easy*

I have an addiction. Well, maybe I have a few addictions. But one of them is definitely Pinterest. It's got something for everyone! It's a virtual feast for the eyes, and I've found it to be a great source of inspiration. When a little can cake pin popped up in front of me, I was intrigued. Baking cakes in used (but clean) tin cans—how brilliant is that? Turns out my grandma had done it since she was a young child in her mother's kitchen!

Well, you know I wanted to put my own spin on it.

I made three can cakes in various sizes and then covered them in a soft-pink glaze to make them look like pillar candles, then placed them on a brownie cake stuffed with little white hearts.

When I presented the cake to my sweet husband, he thought I had just put some big candles on the cake. When I took a fork and cut into the candle, he was shocked! I would pay money to see that expression on his face again. It takes a lot to surprise him!

pink glaze

4 cups powdered sugar

½ cup milk

2 tablespoons light corn syrup

2 to 3 drops pink gel food coloring

Sift the powdered sugar to remove any lumps. Mix all the ingredients into a bowl until combined. Add more milk as needed to achieve the desired consistency.

Baking

1. You'll need 3 aluminum cans of different sizes. For this cake I used a 6-ounce, an 8-ounce, and a 24-ounce can. Clean the cans thoroughly and make sure all the paper is removed. Then dry them carefully and spray the interiors with nonstick baking spray, being sure to coat the bottom and sides.

2. Fill each can half full of white cake batter.

3. Pour the remaining white cake batter into a prepared 8-inch round cake pan.

4. Place the can cakes on a small baking sheet to hold them steady and bake them at 350°F. Set the 8-inch cake directly on the oven rack next to the baking sheet. I started checking on my cakes at 15 minutes and found the smallest tin can cake to be almost done.

5. Remove each can cake when the top springs back or an inserted skewer or toothpick is removed clean. My cakes looked like this directly out of the oven. Cool the cakes to room temperature.

6. Remove the cakes from their cans and slice off the tops so that they're level on both ends. Set the cakes on a wire rack placed over a clean baking sheet. (You can cover the baking sheet with parchment to help with cleanup.) Freeze the can cakes for about 15 minutes.

8. Using an offset spatula, smooth the glaze over the sides of all the cakes, making sure they're totally covered. Let this layer dry completely (you can freeze them for 1 hour or let them harden in the fridge overnight if you're not finishing the same day). Cover the remaining glaze with a wet towel or plastic wrap to keep it from drying out.

9. Repeat with a second layer of glaze. I found that two applications was enough, but you can repeat as many times as needed to achieve the look you want.

Making the Surprise

10. Prepare the buttercream.

11. Remove any hard bits and crumble the remaining white layer into a large bowl. Add 2 to 3 tablespoons of buttercream and combine to make a cake mixture (see page 11). You're looking for a Play-Doh-like consistency.

7. Make the pink glaze. Pour a spoonful of glaze over each can cake, letting it flow down the sides.

12. Roll out a small ball of cake mixture in your hand and shape it into a little cone about 1 inch tall. Repeat, making about 20 total.

13. Insert a toothpick into the base of each cone and move it around in a circle to create a small hole about ½-inch deep. You're creating a three-dimensional heart!

14. Make the brownie batter and pour it into a prepared 9-inch round cake pan. Place the white cake hearts in the batter around the inside edge of the pan, point side up. Bake the brownie layer as directed.

15. Cool to room temperature.

16. Flip the brownie layer onto a cake stand. Your cake hearts will now be right side up when you cut into the cake.

Frosting and Decorating

17. Cover the brownie layer in a crumb coat (see page 13).

18. Insert a coupler into a disposable pastry bag and fill it with buttercream. Attach a 1M tip and pipe roses around the top of the cake (see the Miniature Heart Cake, page 141). Change to the star tip and place small dabs of buttercream around the base of the cake. (I placed a small amount of pink buttercream in my pastry bag when piping out the dabs on the side of the cake.)

19. Gently place the candle cakes on top of the roses, with the largest in back and the smallest in front. Insert a birthday candle into the top of each candle cake so that it's just barely sticking out.

20. Immediately before serving, light the candles. Enjoy!

kiss cake

As you can imagine, my family gets to hear lots about surprise-inside cakes, and I'm forever asking them if they have fun or new ideas for me. My brother-in-law Kurt suggested the Canadian flag. While I love Canada, I think I would give myself an aneurysm trying to figure out how to design a maple leaf inside a cake. But I'm willing to try at some point! Maybe after all my kids have graduated?

My sister had a more doable suggestion: a kiss cake—smooches, not the band. I thought it was a brilliant idea. I would dedicate this cake to her, but that might be kind of weird.

2 recipes White Cake,
page 19

1 recipe Red Velvet Cake,
page 21

1 recipe Basic Buttercream,
page 24

Red gel food coloring

SPECIAL EQUIPMENT:
*4 8-inch round cake pans;
9 × 13-inch pan; offset spatula;
4½-inch round cookie cutter;
sharp, short knife; disposable
plastic pastry bags;
#7 decorating tip*

DIFFICULTY: *Medium*

Baking

1. Bake 4 layers of white cake in 8-inch round cake pans. Cool the layers to room temperature.

2. Bake the red velvet cake in a 9 × 13-inch pan. Cool to room temperature.

3. Level each white layer (see page 9).

4. Prepare the buttercream.

5. Place one white cake layer on a sheet of parchment. Using an offset spatula, cover the top of the cake with ½ cup buttercream, spreading it evenly (see page 13). Place a second layer of white cake on top.

6. Repeat with the other 2 layers. You now have two 2-layer cakes.

7. Freeze both cakes for at least 6 hours, preferably overnight. The cakes should be firm, not crumbly.

Making the Surprise

8. Center a 4½-inch round cookie cutter on top of one of the cakes and gently press down to make a guideline. Repeat with the second cake.

9. Using a sharp, short knife, carve a shallow V shape into the first cake, following the guideline. This will be the lower lip, or the bottom of the kiss cake. Place the cake scrap in a bowl and set this cake aside.

10. Working on the other cake, cut a rounded channel with an island in the middle, inserting the knife about 1 inch into the center of the cake, pointing outward at a 45-degree angle. Make a cut around the entire cake, then move the knife out to the guideline and cut inward at a 45-degree angle from the other side of the channel.

11. Remove the doughnut-shaped cake scrap and set it in the bowl with the other cake scrap.

12. Since an upper lip is generally a smooth, rounded shape, use a spoon to smooth out the valley of your channel. The center island peak can remain sharp and crisp.

13. Next, create a cake mixture (see page 11). Remove any burned or hard edges from the red velvet cake and crumble it into a bowl. Measure 1 cup of the crumbled cake and place into a small bowl. Use your hands or a fork to mix in 2 to

4 tablespoons of buttercream—enough to give you a slightly wet and pliable consistency, like Play-Doh.

14. Add a small dollop of buttercream on the edge of the bottom-lip layer cake. Then add enough red cake mixture to fill the wells in both layer cakes. Freeze any remaining red velvet cake mixture in an airtight container or plastic bag.

15. Spread a thin layer of buttercream around the top perimeter of the bottom-lip cake, avoiding the red velvet cake mixture.

16. Place a small scrap of white cake on the red cake mixture on the bottom cake. In the finished cake, this will show where the lips are parted.

17. Gently invert the upper-lip cake and place it on top of the bottom-lip cake.

18. Cover the cake in a crumb coat (see page 13) and refrigerate for at least 1 hour.

Frosting and Decorating

19. Cover the cake in a smooth layer of white buttercream (see page 14).

20. Tint the remaining buttercream red. Insert a #7 decorating tip into a disposable pastry bag and fill it with the red buttercream. Pipe puckered lips randomly over the cake. Then pipe out a base around the cake.

21. Chill the cake until serving.

miniature heart cake

2 recipes White Cake, page 19

1½ recipes Basic Buttercream, page 24

Red gel food coloring

SPECIAL EQUIPMENT:
2 9 × 13-inch cake pans; ruler; long serrated knife; disposable plastic pastry bags; 1M decorating tip

DIFFICULTY: *Medium*

I really wanted to make a cake that would be visually stunning and easy to decorate. Why? Because I want you to be able to get maximum impact for the least amount of effort. Like me, you're busy. You're pulled in multiple directions, whether you're a mom or dad or daughter or friend.

But I also know that you love to give and share with others. While we often have the grandest of intentions when it comes to giving, sometimes life just steps in the way. Even though you want to bake cupcakes for your neighbor, your son's class needs them, too. You may want to make a cake for your local fire department, but you have six family birthdays to celebrate this month alone. Sometimes it's just hard.

You can make this cake in one afternoon or over the course of a couple days, but the decorating you can do in a matter of minutes! From beginning to end this frosting rose technique that I created took less than 10 minutes. How about that for maximum impact with minimal effort?

Baking

1. Bake 2 white cakes in 9 × 13-inch rectangle pans. Cool to room temperature and place the cakes on parchment.

Making the Surprise

2. Make 1 batch of buttercream. While the buttercream is still in the mixer, add red food coloring. I used about 1 tablespoon to reach my desired red. Place the red buttercream in the fridge until you need it.

3. Make ½ batch of buttercream and leave it white.

4. Start by carving the edges off of the top layer (above at left), cutting at an angle so that the base of the top layer stays as wide as the bottom layer but the top has more of a domed or trapezoid shape. Save all the cake scraps in a bowl.

5. Trim the top and sides of the bottom cake layer. Using a ruler, measure in 1 inch from all sides and cut a guideline around the entire bottom layer. You can also take a piece of paper that is 1 inch smaller than the cake's dimensions and use it as a template.

6. Cut a small inverted V on the guideline by inserting your knife into the cake at a 45-degree angle just ¼ inch or so outside the guideline and making a cut along the whole cake. Finish the V

by inserting the knife at a 45-degree angle ¼ inch on the other side of the guideline, meeting the other cut, and cutting around the cake. Remove the excess cake from the V and save it with the other scraps.

7. Make a red cake mixture (see page 11) by adding about 1 tablespoon buttercream and about ½ teaspoon of red gel food coloring to the white cake scraps. Mix until the cake mixture is fully incorporated and pliable. Add more red if you want a deeper color.

8. Working on parchment so that the red doesn't dye the surface of your counter, roll out 4 "snakes" to fill the 4 sides of the V channel in your cake. (Mine was about ¼ inch in diameter.) Make 2 snakes about 7 inches long and 2 about 11 inches long.

9. Gently press down one side of each snake to make the point at the bottom of the heart.

10. Carefully fit the snakes into the V channel placing the point toward the bottom. If they break, gently press them back together.

11. Run the dull side of a butter knife down the center of each snake. This will be the indentation at the top of the heart.

12. Spread the white buttercream around the sides of the bottom cake.

13. Gently place the top layer of the cake onto the bottom layer. (You can also brush the bottom layer with about 1 tablespoon of Simple Syrup, page 279, to help the layers stick together.) Cover the top of the cake in a thin crumb coat of red buttercream (see page 13). This will help to create a seamless look when you pipe the roses on.

14. Freeze the cakes for 1 hour or refrigerate them for a few hours.

Frosting and Decorating

15. The rose buttercream technique uses a lot of frosting, so fill three pastry bags with red buttercream (one full batch of my buttercream recipe) and seal the tops with a rubber band or twist tie. In a separate bag, insert a 1M tip. Cut off the tip of one of the prepared pastry bags and insert it into the bag with the 1M tip. When you run out of frosting, just pull out the empty bag and insert the next full bag.

16. To make a rose, apply pressure to the bag and make a center dollop. Slowly move around the dollop in a circle. I went around each center twice, but you can also just go around once for a tighter bud. Continue around the entire cake. If you have any dead space, go back in and add a small swoop. Try not to dab or make stars.

17. Chill the cake until serving.

sunset cake

Sunsets rock. (I'm sure sunrises are amazing, too, but Mama loves her sleep!)

I wanted to create a really simple cake that would transport you—that would instantly remind you of a happy time and place where you experienced nature's beauty. Hopefully you were on a beach somewhere, with someone you love.

Now that I think about it, this cake might need a bit more research. I'd better let hubby know we need to head to a tropical island, pronto.

Baking the Surprise

1. Place ¼ cup of the white cake batter in a small bowl and add a drop or two of yellow food coloring. You're looking to achieve the color of the setting sun. Pour the yellow batter into a cupcake tin.

2. Divide the remaining batter among 4 bowls, roughly 1 cup per bowl.

3. Add 1 drop of orange to one bowl and 3 to 5 drops to another bowl. Blend thoroughly.

4. Add 1 drop of turquoise gel food coloring to a third bowl and 3 to 5 drops to the last bowl. Blend the batters thoroughly.

5. Pour each batter into a separate disposable pastry bag.

1 recipe White Cake, page 19

1 recipe Basic Buttercream, page 24

Yellow, orange, and turquoise gel food coloring

SPECIAL EQUIPMENT:
2 8-inch round cake pans; cupcake tin; rubber spatula; disposable plastic pastry bags; cake leveler or long serrated knife; 127D decorating tip; small offset spatula

DIFFICULTY: *Medium*

6. Prepare the pans by spraying with baking spray, then place a piece of parchment on the bottom.

7. Cut off a small tip of the pastry bag with the darkest blue cake batter and pour some into the prepared cake pan. Spread a thin layer around the bottom of the pan.

8. Cut the tip off the light blue pastry bag and pipe out some batter onto one side of the cake pan.

9. I tried to imagine I was creating a wave effect with the batter, so I put a majority of the light blue on one side, then filled in the other side with the dark blue.

10. Then just fill in with all of the remaining blue batters. You really can't do it wrong as long as you don't start mixing the batters!

11. Spread a thin layer of the darker orange batter on the bottom of the pan.

12. Make a little doughnut around the middle of the cake with the lighter orange batter. Later, you'll be carving out the center to hold the yellow sun.

13. Add more dark orange batter, piping it over the light orange doughnut and around the edges of the pan. Then fill in the rest of the pan with all the remaining light and dark orange batters.

14. Bake the cakes as you normally would. Bake the yellow cupcake at the same time. The cakes may not come out perfectly flat, and that's fine.

15. Cool the cakes to room temperature, then freeze the orange layer for at least 30 minutes.

16. Prepare the buttercream and divide it evenly among 3 bowls. Tint one bowl yellow, one the same blue as the blue layer, and the remaining bowl the same orange as the orange layer.

17. Place the yellow cupcake on the center of the chilled orange layer. This will give you an idea about how deep and wide to carve.

18. Using a soupspoon, begin to gently remove some of the center of the orange layer. When you've created a small indentation, place the cupcake in the hole and determine if you need to cut out more cake. The cupcake should sit about level with the top of the orange layer.

19. Place the cupcake snugly into the orange layer and refrigerate for 15 minutes.

20. Place the leveled blue layer on the cake stand. Cover the top in about ½ cup of blue frosting (see page 12).

21. Carefully flip over the orange layer onto the blue layer.

Frosting and Decorating

22. Cover the top of the cake in orange buttercream. Try to make the top as smooth as you can, but it doesn't need to be perfect.

23. Place the remaining blue, orange, and yellow buttercreams in individual pastry bags.

24. Place a 127D tip into an empty pastry bag. Cut off the excess bag so that the tip fits snugly in the bag. This will allow you to use the same tip quickly with the three different colors. You will not need to clean the tip when you change out the bags.

25. Cut a large amount off the tip of the yellow frosting bag and drop it into the bag fit with the 127D tip. Center the tip between the top and base of the cake and, holding your tip flat against the side, pipe out a ribbon of yellow buttercream around the entire cake. You want a consistent line, but it doesn't need to be perfect.

26. Remove the yellow frosting bag and insert the blue one. Pipe out a bit of buttercream to make sure all the yellow buttercream is out, then pipe out a ribbon of blue buttercream around the bottom of the cake, overlapping the yellow.

28. Hold a small offset spatula perpendicular to the cake. Gently and carefully start to pull the knife around the cake. You can go around the cake as many times as you want to achieve the desired sunset effect.

29. Clean the offset spatula and use it to carefully smooth the top edge of the cake.

30. Chill the cake until serving.

27. Remove the blue frosting bag and insert the orange one. Pipe out a bit to make sure all the blue buttercream is out, then pipe out a ribbon of orange buttercream around the top of the cake.

ring cake

2 recipes White Cake,
page 19

1 recipe Basic Buttercream,
page 24

Black, white, and gold
(or yellow) gel food coloring

1 small jar of white
sparkling sugar

SPECIAL EQUIPMENT:
*4 6-inch round cake pans; small,
sharp knife; rotating cake stand
(optional); rolling pin; 2-inch
round cookie cutter; 3-inch
round cookie cutter; offset spatula*

DIFFICULTY: *Challenging*

Have you ever been to an engagement party? They sound absolutely fabulous. Like a wedding party with all the same people but less formal and a lot more fun.

I didn't have an engagement party. The day I got engaged I went out to Tailgator's Sports Bar in Fargo, North Dakota. We drank cold beer and played a few games of pool, and we had a fabulous time—just me, my fiancé (I must have said that word a hundred times that night), my sister, and her hubby. I wore a permagrin all night long.

Just in case *you* have an engagement celebration, this little cake is a perfect way to introduce your friends and family to your new bling!

Ring shown actual size. (Totally kidding.)

Baking

1. Prepare the batter for the white cakes. Remove 2 cups and set it aside in a small bowl. Add a couple drops of black food coloring to the remaining 6 cups of the batter to make it gray. If you like, you can make the 2 cups of batter a brighter white by adding white food coloring.

2. Bake 3 layers of gray cake and 1 layer of white cake in 6-inch pans. (You may need to extend the recommended cooking time when using 6-inch pans). Cool the cakes to room temperature, then freeze the gray layers for no less than 6 hours, but preferably overnight.

3. Prepare the buttercream.

4. Remove any hard bits of crust from the 6-inch white cake and crumble three-quarters of it into a large bowl.

5. Mix in 2 teaspoons of the buttercream and 1 or 2 drops of white gel food coloring, as needed to reach the desired color.

6. In a separate bowl, crumble the remaining white cake and combine it with about 1 teaspoon of buttercream and a drop or two of yellow or gold food coloring, to approximate the color of a gold ring.

Making the Surprise

7. To make the diamond portion of the cake, place a gray layer flat on parchment paper (using a rotating cake stand here is a great idea). Mark the center of your cake with a toothpick.

8. Insert a small, sharp knife at a 45-degree angle about halfway between the toothpick and the edge of the cake. Aim for the toothpick as you cut. Once you hit the toothpick or the bottom of the cake, start slowly moving the knife around the cake in a circle to cut out the shape of a cone. If you're using a rotating cake stand, use one hand to guide the knife and the other to slowly turn the cake.

9. Pick up the toothpick to remove the cake cone. Set it aside in a safe place—you'll need it later.

10. Lay the remaining gray layers on a sheet of parchment paper. Center the cookie cutter on top of one cake layer and gently press a circle as a guideline for where to cut. Repeat on the other gray layer. I thought it would be fun to have an exaggerated diamond on a small ring, so I chose a 2-inch round cookie cutter for my cake.

11. Use a soupspoon (or any small spoon) to cut out circular shapes. Move the spoon around the line created by the cookie cutter and cut a half-moon shape from the cake. If you don't scoop out enough the first time, or if there is an abundance of crumbs, simply clear out more cake. Reserve the cut-out cake in a small bowl. If your cake seems crumbly or too soft, return it to the freezer for 15 minutes.

12. Repeat to carve a well in the second gray layer, making sure it matches the first.

14. Use a 3-inch round cookie cutter to cut out 2 circles, then make a slit halfway through.

13. Using a sheet of parchment, roll out a thin layer of the gold cake mixture.

15. Gently place the circle into the cavity of one of the gray layers, filling the entire space. Use a knife to remove any excess gold cake mixture. Repeat with the other gold circle and gray layer.

16. Mix a scant drop of buttercream into the reserved gray cake to make cake mixture.

Note: *The cone of cake on the toothpick should still be reserved separately.*

17. Fill both gold cups with the gray cake mixture. Pack it carefully, but really get it in there—if you do not use enough, the ring could collapse when you cut the cake to serve it.

18. Place 1 cup of buttercream in a separate bowl and tint it gray to match the cake using a drop or two of black gel food coloring.

19. Using an offset spatula, spread half the gray frosting around the top perimeter of one of the gold ring layers (don't get any frosting on the gold or inner gray parts). Turn over the second gold ring layer and lay it carefully on top, so that the ring sides touch each other. Go slowly and make sure the ring parts match!

20. After icing the last layer, set the final single layer on top of the cake and freeze the cake while you work on the diamond.

21. Mix a couple of tablespoons of sparkling sugar into the white cake mixture. The sugar seems to lose a little of its sparkle when mixed into the cake mixture, but I like the overall effect!

22. To make the diamond, look at the cone of cake you removed with the toothpick and make a white sparkly cone to match that shape and size. Place the diamond into the cavity on top of the cake and see how much it fills the space. It should not go as high as the top of the cake, so cut off enough to leave ½ inch or so of space.

23. Remove the toothpick from the gray cake cone and slice off ½ inch or so from the top. Place this cap on top of the diamond. Gently cover the top of the cake in gray buttercream (see page 12).

24. Crumb-coat the whole cake with white buttercream (see page 13) and chill for at least 1 hour in the refrigerator or 15 minutes in the freezer.

Frosting and Decorating

25. Place the cake on a cake stand. Frost the cake with a smooth layer of white buttercream (see page 14), focusing on making the sides very smooth. Leave the top rough. Dust the top of the cake with sparkling sugar, taking care not to spill it over the sides.

26. Place the cake stand on a clean baking sheet. Pour sparkling sugar into your hand and gently press it into the lower part of the cake. You can reuse the sugar that falls onto the baking sheet.

27. Chill the cake until serving.

CELEBRATING HOLIDAYS

nothing inspires me more than creating a cake for a holiday. I get excited thinking about people all across the country celebrating something at the same time. It's like that scene in *Ghostbusters II* where all of New York stops and hugs one another and sings together.

Okay, so maybe it's not really like that. But that's a good movie, right?

new year's eve cake

I'm not kidding when I tell you that people are going to flip for this cake, which consists of thirteen layers of chocolate and Champagne buttercream deliciousness. It was perfect at our 2013 New Year's Eve celebration! What's the best part? It can easily be modified for any year New Year's party...just use a number of layers to match the year: 2025 should be pretty interesting!

3 recipes Chocolate Cake, page 20

2 recipes Basic Buttercream, page 24, substituting Champagne for the milk

Gold sprinkles

SPECIAL EQUIPMENT: *6 9-inch round cake pans; cake leveler or long, serrated knife; small and large offset spatulas*

DIFFICULTY: *Easy*

Baking

1. Bake 6 chocolate cake layers in 9-inch round cake pans. (You may not need this many cakes, depending on how many layers you ultimately want. Each full layer will make 3 thin layers.) Cool the layers completely, then freeze for at least 1 hour. The cakes should be very firm but still able to be cut.

2. Prepare the Champagne buttercream.

Making the Surprise

3. Using a cake leveler, slice each cake into 3 thin layers, each about ½ inch thick. You may have layers left over, depending on the year.

4. Place 1 layer on a cake stand and, using an offset spatula, spread the top with ¼ cup of buttercream. Smooth the buttercream as evenly as you can—some bakers will actually use a carpenter's level during this process. Just be sure to step back and look at your cake periodically to make sure it's level. If one side is too high, overcompensate with buttercream on the next layer to create a flat, even surface. If you find

that your thin layers are getting too crumbly at any point, just place them back in the freezer to firm up before you finish the cake.

Frosting and Decorating

5. If your cake is steady and firm, cover it in a crumb coat (see page 13). Refrigerate for at least 1 hour.

6. Cover the cake in buttercream, making the sides very smooth and leaving the top rough. Sprinkle the top generously with gold sprinkles.

7. Chill the cake until serving.

tips and tricks

This cake is tall and makes very wide servings! Serve slices on dinner plates, not cake plates.

christmas tree cakes

Christmas inspires me. I love the energy and excitement that December holds! There are so many fun occasions and parties and reasons to joyfully celebrate, and this cake is perfect for all of them!

This recipe makes two 3-layer cakes.

4 recipes White Cake, page 19

2 recipes Basic Buttercream, page 24

Green, red, yellow, and blue gel food coloring

SPECIAL EQUIPMENT:
6 8-inch round cake pans; cake leveler or long serrated knife; paring knife; small offset spatula; disposable plastic pastry bags; #4 and #2 decorating tips

DIFFICULTY: *Medium*

Baking

1. Evenly divide 2 recipes of white cake batter among three 8-inch round cake pans. The easiest way to do this is to double a white cake recipe and prepare it in a stand mixer. You'll use about 2¾ cups of batter per pan.

2. Bake, then cool the layers to room temperature. You may need to bake them for 5 to 10 minutes longer than normal.

3. Prepare another double batch of white cake, this time tinting the batter green (see page 11). I chose forest green gel food coloring, but you can pick whatever green you like. Divide the batter among 3 pans, bake, and cool to room temperature.

4. Identify the shortest cake of the six, and level the other five cakes to the same height. Freeze the cakes for at least six hours, preferably overnight.

Making the Surprise

5. Trace out 3 circles onto parchment or wax paper, one 6 inches in diameter, one 4, and one 2. (You can also use cookie cutters or any household objects with similar dimensions.) Cut out the circles.

7. Swap out the cut parts so that the green insert is in the white cake and the white insert is in the green cake.

8. Repeat steps 6 and 7 with the 4-inch round circle on another pair of white and green layers. Swap out the centers.

9. Place the 2-inch parchment round in the center of the last white layer. Using a toothpick or skinny skewer, find the center of the cake and insert it through the parchment round and all the way through the cake. Insert the knife at the guideline at a 45-degree angle and cut around the layer, making sure the tip of the knife is touching the toothpick. You will have cut out a cone shape. Remove the toothpick and parchment round and repeat on the green cake.

10. Swap the cones, placing the white cone in the green cake and vice versa.

11. As you build the tree cakes, note that all the layers will have to be turned upside down to mimic a tree shape. Working first on the green cake with the white tree, place the first layer you created (with the 6-inch cutout) on the cake stand, upside down.

6. Place a white layer on a sheet of parchment paper. Center the 6-inch parchment round on top of the cake. Then insert a sharp paring knife into the layer at about a 60-degree (very sharp) angle, pointing toward the center of the cake, and make an even cut all the way around. Repeat the same cut on one of the green layers.

12. Prepare one recipe of buttercream and color it the same green as the cake. Prepare the second batch of buttercream, leaving it white.

13. Using a small offset spatula, spread green buttercream on the top of the green part of the layer and white buttercream on the white tree part. Top with the second green layer (with the 4-inch cutout) and frost the same way, then add the third layer.

14. Cover the cake in a crumb coat of green buttercream (see page 13).

15. Repeat to make the white cake with the green tree. Refrigerate both cakes until ready to decorate.

Frosting and Decorating

16. To decorate the white cake, cover the cake in white buttercream, smoothing out the sides but leaving the top rough (see page 12).

17. Insert a #4 decorating tip into a disposable pastry bag and fill it with green buttercream. Carefully pipe out a garland around the top of the cake.

18. Tint ¼ cup of the white buttercream red, ¼ cup yellow, and ¼ cup blue. Then fill pastry bags with each color.

19. Insert a #2 decorating tip into a pastry bag then cut the tip of the bag of red buttercream and slip it into the bag with the tip. Pipe little Christmas lights onto the garland you just created. Switch out the red frosting for the yellow, pipe out a bit of buttercream to clear out any red frosting still in the tip, and pipe more Christmas lights. Repeat with the blue buttercream.

20. Frost the green cake the same way, but use only white buttercream.

21. Chill the cakes until serving.

cross cake

SPECIAL EQUIPMENT:
*4 8-inch round cake pans;
9 × 13-inch pan; 2-inch and
4½-inch round cookie cutters;
offset spatula.*

DIFFICULTY: *Medium*

I loved this cake from the moment of its conception. It's simple but so effective in its theme. I also love my Thursday morning Bible study. Our church has a great fondness for Beth Moore, and our group has completed quite a few of her amazing studies.

Well, one day it occurred to me that I needed to send Beth a picture of my cross cake. I felt the blood rushing through my veins as I thought about what would happen if I managed to get her a cake in person. I could envision it so clearly—she would see this cake and embrace me and declare me her long-lost BFF! We would dance in circles around her office and collapse into a fit of giggles that only the best of friends share.

So I rushed home and e-mailed her a photo through her website. I waited a few days and heard nothing. How odd! So I e-mailed again. What if the people on her staff didn't realize that this cake was going to change Beth's life?!

Two years have passed and I don't think I'm going to hear from Beth. I think now that I may have come across as a little stalkerish. It can't be the cake—the cake is good. My stalkerlike tendencies? Well, I guess they could use some work.

Baking

1. Bake 4 white layers in 8-inch round cake pans. Cool the layers to room temperature, then freeze them for about 1 hour, or until firm but not frozen.

2. Bake the chocolate cake in a 9 × 13-inch pan.

3. Prepare the buttercream.

Making the Surprise

4. Stack 3 layers of white cake on a piece of parchment paper—no buttercream between the layers, just a stack of cakes.

7. Place the fourth white layer on a sheet of parchment and center a 4½-inch round cookie cutter on the layer.

5. Using a 2-inch round cookie cutter on the center of the cake, push through all 3 layers, removing the excess white cake as you go. Set the cakes aside.

6. Press the same cutter into the chocolate cake to make 3 chocolate cake columns. Set them aside.

8. Press to cut through the cake, then remove the cut center and set it aside.

9. Press the same cutter into the chocolate cake to cut out a disk.

10. Place one of the layers with a 2-inch hole on a cake stand. Using an offset spatula, place a thin layer of buttercream on top. Then add another layer and top that layer with buttercream as well.

11. Insert 2 of the chocolate cake columns into the center hole. If they're too tall, simply trim off the excess chocolate cake.

12. Tint 1 cup of the buttercream brown to match the chocolate cake and spread a bit of it on the brown column.

13. Top the cake with the large cutout white layer circle. Insert the 4½-inch chocolate circle in the middle. Then cover the white part of this layer with white buttercream and the chocolate part with brown buttercream.

Frosting and Decorating

16. Cover the cake in a thick layer of buttercream (see page 12).

17. Using a wooden spoon or other utensil with a sturdy rounded tip, create swoops in the buttercream. If you like, use a toothpick to create your initial design, then trace it with the blunt, rounded edge of the spoon.

14. Top with the last white layer and insert the last chocolate column.

15. Cover the cake in a crumb coat (see page 13) and freeze for at least 3 hours to overnight.

hanukkah cake

There's something so beautiful about the colors and symbols associated with Hanukkah—the deep blues, the perfect whites, the menorah. I couldn't help but be inspired by the beauty of it all. And you know what I do when I get inspired . . . I bake a cake!

3 recipes White Cake, page 19

1 recipe Basic Buttercream, page 24

Royal blue gel food coloring

SPECIAL EQUIPMENT: *4 8-inch round cake pans; 9 × 13-inch cake pan; 2-inch and 4½-inch round cookie cutters; paring knife; rotating cake stand; small and large offset spatulas; disposable plastic pastry bags; #4 decorating tip.*

DIFFICULTY: *Medium*

Baking

1. Bake 2 recipes of white cake in 4 8-inch round cake pans.

2. For the third white cake recipe, add 4 to 10 drops of royal blue gel food coloring to the batter, as needed to make it bright blue. Bake it in a 9 × 13-inch pan.

3. Cool the cakes to room temperature, then freeze them for at least 6 hours.

4. Prepare the buttercream.

5. Crumble the blue cake into a large bowl, removing any dark or hard edges. Add 2 to 4 tablespoons of buttercream and combine to make a smooth cake mixture (see page 11). You're looking for Play-Doh-like consistency. Place the cake mixture in a plastic bag or cover the bowl in plastic until you're ready to assemble the cake.

Making the Surprise

6. Place 2 white layers on a piece of parchment paper. Center a 2-inch round cookie cutter on one layer and press it in slightly to create a guideline. Repeat on the other layer.

center of the layer; stay about 1 inch away from it on all sides. Work your way around the layer, following the guideline, then remove the piece of cake from the center; it should be 4½ inches wide on top and 2 inches wide on the bottom. Repeat on the other layer.

7. Insert a sharp paring knife into the guideline at a 45-degree angle and cut toward the center of the cake. Work your way around the cake, following the guideline, to cut out a cone. This works well if you have a turntable or rotating cake stand. Remove the cone of cake and set it aside. Repeat this on the other layer.

8. Place the 2 remaining white layers on a piece of parchment.

9. Center a 4½-inch round cookie cutter on one of the layers and press in slightly to create a guideline. Repeat on the other layer.

10. Insert the knife at about a 45-degree angle on the guideline pointed toward the center of the cake. Don't press the knife all the way to the

11. To assemble the cake, place one of the layers with the small cone cut out on a cake stand. The point of the cone should be on the bottom.

12. Mold some blue cake mixture so that it will fill the cone shape and press it into place. Adjust as needed.

13. Fill all the remaining layers with cake mixture, molding it into the voids and adjusting as needed.

14. Using an offset spatula, place a thin layer of buttercream on top of the bottom layer, taking care not to get any on the blue part. Then carefully flip one of the 4½-inch cut layers and set it on top (the wider blue part should face down).

15. Spread another thin layer of buttercream on top of the white cake, taking care not to get any on the blue cake mixture.

16. Top with the second 4½-inch cut layer, this time with the wider blue side facing up. Carefully match up the edges of the layers, as this is where the Star of David meets in the middle.

17. Spread another thin layer of buttercream on the white cake, taking care not to get any on the blue cake mixture. Then add the last layer, flipping it so that the point of the cone is facing up this time.

18. Cover the cake in a crumb coat (see page 13) and freeze for 15 minutes if necessary.

Frosting and Decorating

19. Cover the cake in a smooth coat of white buttercream (see page 12). Tint the remaining buttercream a deep blue to match the interior of the cake.

20. Insert a #4 decorating tip into a disposable pastry bag and fill it with the blue buttercream. Pipe out a scroll pattern around the entire cake. You can practice the pattern by drawing it on a sheet of paper and laying parchment on the top; pipe right onto the pattern. Or, if you prefer, you can use a toothpick to gently trace the design on the cake before you pipe it with buttercream.

21. Chill the cake until serving.

red velvet holiday candle cake

SPECIAL EQUIPMENT:
*4 8-inch round cake pans;
cupcake tin; offset spatula;
2-inch round cookie cutter; #102
and #4 decorating tips.*

DIFFICULTY: *Medium*

I don't tell people this often, but I secretly wish I'd been born in the South. In the late 1980s, when *Designing Women* was popular, I was just entering my formative teen years. I would watch Julia Sugarbaker and dream of being a graceful, witty, charming Southern girl instead of the chubby, insecure, socially awkward girl I was.

Fast-forward twenty years, and one of my favorite magazines is *Southern Living*. I soak it up! When I heard they were having a contest to have a reader's cake featured on the cover, I nearly fell off my chair. I thought about it day and night. What could I submit? What would a Southern reader appreciate in a cake? And it dawned on me . . . what flavor does every Southerner love? Red velvet.

So I reinvented my Jack-o'-lantern Cake (page 189) into a Red Velvet Holiday Candle Cake. I made the cake, took pictures, and submitted my recipe for shortening-based cream cheese frosting.

Southern Living did not pick this Minnesota girl. But their prompting helped me discover another way to enjoy the candle surprise-inside cake! And I have to admit, it's one of my favorites.

So I do believe I owe a big thank-you to those fabulous folks!

Baking

1. Bake the red velvet cake into four total 8-inch layers. Cool them to room temperature.

2. Prepare the white cake batter. Reserve ½ cup of the batter and stir in 1 to 3 drops of yellow gel food coloring. Reserve another ¼ cup of the batter and stir in 2 to 4 drops of orange gel food coloring.

3. Pour half the yellow batter into one well of a prepared cupcake pan. Pour about 1 tablespoon of the orange batter right over it. Then repeat to create a second yellow-and-orange cupcake. You only need one to make the candlestick flame, but it's best to make two in case one doesn't work out.

4. Insert a toothpick into the center of the orange/yellow batter a few times. This will help to mix the colors and create the appearance of a flame.

5. Pour the white cupcake batter into the remaining cupcake wells and bake as directed. Chill the cupcakes in the refrigerator for 1 hour or in the freezer for at least 10 minutes. (You'll have plenty of extra cupcakes to enjoy.)

6. Prepare the buttercream. Tint 2 cups a deep red to match the color of the cake.

Making the Surprise

7. Place 1 cake layer on a piece of parchment and, using an offset spatula, cover the top with a thin layer of red buttercream (about 1 cup). Add a second layer and another thin layer of the red buttercream. Place a third layer on top and freeze the cake for at least 6 hours, preferably overnight.

8. Place the 3-layer red velvet cake on a sheet of parchment. Grab a 2-inch round cookie cutter (or any round cookie cutter that's smaller than the base of your cupcake) and center it on your cake. Then push the cutter through all the cake layers, removing the excess cake as you go. Just push on through—don't worry about smushing the cake within the cutter!

9. Using a small paring knife, carve out a little hole in the fourth layer of red velvet cake. The hole should be smaller than the 2-inch cookie cutter, as this is where you will be putting the orange-yellow flame.

10. Unwrap 3 of the white cupcakes and, using the same (cleaned) cookie cutter, create 3 round white cake discs. To create a seamless transition and a beautiful candle, remove all the hard edges and the darker tops and bottoms of the cupcakes. This is easiest to do when the cupcakes have been chilled.

11. Carefully place the 3 white cupcake layers one at a time in the center of the red velvet cake. Make sure the bottom cupcake gets all the way to the bottom (you can gently pick up the cake and look at the bottom to verify). If you have any excess white cake sticking out, simply trim it to the same level as the top of your cake. Now your candlestick is in place!

12. Refrigerate the cake while you create the flame.

13. Carve around the edge of one of the orange-yellow cupcakes until you have a flame shape that will fit into the hole you carved in the fourth layer. (If you don't like the looks of that flame, go to your backup flame cupcake!) Insert the flame into the fourth layer. The flame design is very forgiving, so don't be afraid to work it in there.

14. Remove the 3-layer cake from the refrigerator and place it on a cake stand. Spread a thin layer of red buttercream around the top of the cake, taking care to not get any on the white candlestick. Set the fourth cake layer on top, flame side down.

15. Cover the cake in a white crumb coat (see page 13). Place the cake in the freezer for 30 minutes or in the refrigerator for a few hours.

Frosting and Decorating

16. Cover the top of the cake in a thin layer of white buttercream. The sides will be covered by the design, but the top will remain undecorated, so make sure it is as clean and unblemished as possible (see page 12).

17. Insert a #102 decorating tip into a disposable pastry bag and fill it with white buttercream. Starting at the bottom of the cake, with the small point in the tip facing out, apply gentle pressure to the bag. Move your bag back and forth up the side of the cake until you reach the top.

18. Insert a #4 tip into a disposable pastry bag and fill it with red buttercream. Starting at the top of the cake, make a small dab of frosting, then gently pull it down.

19. Repeat steps 17 and 18, alternating the red and white buttercreams around the entire cake.

20. Chill the cake until serving.

jack-o'-lantern cake

This was my first-ever surprise-inside cake. We typically celebrate Halloween every year at our church's Trunk-n-Treat event; people decorate the back of their vehicles and the kids go around to each car and get candy. Pastor Chris, the church's youth pastor, is a pretty talented baker, so one year I decided to show him that I like to bake, too. I was having trouble thinking of a church-friendly Halloween-themed cake, but I got inspired by the idea of baking a pumpkin cake with a candle *inside*. Thus was the Jack-o'-lantern Cake born!

Truth is, we loved the cake so much it never even made it to church!

2 recipes Chocolate Cake, page 20

1 recipe White Cake, page 19

1 recipe Basic Buttercream, page 24

½ recipe Chocolate Buttercream, page 25

Yellow, orange, green, and black gel food coloring

SPECIAL EQUIPMENT: *4 8-inch round cake pans; cupcake tin; offset spatula; 2-inch round cookie cutter; paring knife; long serrated knife; disposable plastic pastry bags; #4 decorating tip*

DIFFICULTY: *Medium*

Baking

1. Prepare the chocolate cakes. You'll be baking two recipes into four total 8-inch layers.

2. Prepare the white cake batter. Reserve ½ cup of the batter and stir in 1 to 3 drops of yellow gel food coloring. Reserve another ¼ cup of the batter and stir in 2 to 4 drops of orange gel food coloring.

3. Pour half the yellow batter into one well of a prepared cupcake pan. Pour about 1 tablespoon of the orange batter right over it. Then repeat to create a second yellow-and-orange cupcake. You only need one to make the candlestick's flame, but it's best to make two in case one doesn't work.

4. Insert a toothpick into the center of the orange/yellow batter a few times. This will help to mix the colors and create the appearance of a flame.

5. Pour the white cupcake batter into the remaining cupcake wells and bake as directed. Chill the cupcakes in the refrigerator for 1 hour or in the freezer for at least 10 minutes.

Making the Surprise

See Red Velvet Holiday Candle Cake, page 183, for more images with this tutorial!

6. Prepare the chocolate buttercream.

7. When all 4 chocolate cake layers are cooled to room temperature, create a 3-layer cake. Place 1 cake layer on parchment and, using an offset spatula, frost the top with a thin layer of chocolate buttercream (about 1 cup). Add a second cake layer and another thin layer of chocolate buttercream. Place a third layer on top and freeze the cake for at least 6 hours, preferably overnight.

8. Place the 3-layer cake on a sheet parchment. Grab a 2-inch round cookie cutter (or any round cookie cutter that is smaller than the base of your cupcake) and center it on your cake. Then push the cutter through all the cake layers, removing the excess cake as you go. Just push on through—don't worry about smushing the cake within the cutter!

9. Using a small paring knife, carve out a little hole in the fourth layer of chocolate cake. The hole should be smaller than the 2-inch cookie cutter, as this is where you will be putting the orange-yellow cupcake, or flame.

10. Unwrap 3 of the white cupcakes and, using the same (cleaned) cookie cutter, create 3 round white cake discs. To create a seamless transition and a beautiful candle, remove all the hard edges and the darker tops and bottoms of the cupcakes. This is easiest to do when the cupcakes have been chilled.

11. Carefully place the 3 white cupcake layers one at a time in the center of the chocolate cake. Make sure the bottom cupcake gets all the way to the bottom (you can gently pick up the cake and look at the bottom to verify). If you have any excess white cake sticking out, simply trim it to the same level as the top of your cake. Now your candlestick is in place!

12. Refrigerate the cake while you create the flame.

13. Carve around the edge of one of the orange-yellow cupcakes until you have a flame that will fit into the hole you carved in the fourth layer. (If you don't like the looks of that flame, go to your backup flame cupcake!) Insert the flame into the fourth layer. The flame design is very forgiving, so don't be afraid to work it in there.

15. Carve the exterior of the cake so that it resembles the shape of a pumpkin. Holding a serrated knife at a 45-degree angle and starting with a shallow cut, start shaving away until you have created a soft curve.

14. Remove the 3-layer cake from the refrigerator and place it on a piece of parchment paper.

16. You need a nice, round overall shape.

17. Prepare the basic buttercream. Tint 1 cup green and the remaining buttercream orange.

18. Cover the cake in a crumb coat (see page 13) using 1 to 1½ cups of the orange buttercream. Then refrigerate the cake for 1 hour to firm it up.

Frosting and Decorating

19. Carve out another cupcake for the stem of the pumpkin. Place the cupcake on top of the cake and cover it in green buttercream.

20. Apply a smooth layer of orange buttercream to the entire cake. Then, starting from the base of the cake, drag a rubber spatula to the top of the cake and remove the excess frosting while creating the look of the natural lines of a pumpkin.

21. Insert a #4 decorating tip into a disposable pastry bag. Fill a second pastry bag with a small amount of green buttercream. Cut the tip off and slip the bag into the pastry bag with the #4 tip. Pipe some green vines coming down from the stem.

22. Tint any remaining green buttercream black, and fill a pastry bag with it. Slip the bag into the pastry bag with the #4 tip. Pipe out a bit to clear any green frosting. Then pipe a jack-o'-lantern face onto the cake. You can use a toothpick to draw the design on the cake first if you prefer.

23. Chill the cake before serving.

red, white, and blue cakes

SPECIAL EQUIPMENT: *2 8-inch round cake pans; cardboard cake round; small offset spatula; wax or parchment paper; kitchen string; rotating cake stand (optional); donkey and elephant cookie cutters; disposable plastic pastry bags; coupler set (optional); #104 and #4 decorating tips*

DIFFICULTY: *Medium*

When Abraham Lincoln was campaigning for the presidency, he discovered he had made an enemy in Edwin McMasters Stanton. Stanton made a point of trying to embarrass Lincoln publically, often saying degrading things.

After Lincoln won the presidency he had to choose his cabinet. When he was faced with the decision to choose the secretary of war, Lincoln shocked everyone and chose Stanton! People closest to Lincoln were outraged at his choice and said to him, "Mr. President, this is a mistake! Stanton has degraded you and said ugly things about you. He is an enemy and will sabotage you!"

After much deliberation Lincoln replied, "Yes, I know Stanton. But . . . I find he is the best man for the job."

Lincoln's wisdom was evident in his choice. Stanton provided excellent service to his country as secretary of war. After Lincoln was assassinated a grieving Stanton commented, "He now belongs to the ages."

I can't help but be moved by this amazing aspect of history. These words and actions were the inspiration behind the patriotic cake! God bless America!

This recipe makes two 2-layer cakes.

Baking

1. Bake the white cake in two 8-inch round cake pans.

2. While the cakes are still warm from the oven, place them on a flat surface and level them (see page 9).

3. Gently place one layer directly on top of the other, without any frosting in between. When you place warm cakes together and then chill them, it helps to create a seamless look when you cut into it.

4. Repeat to make 2 red velvet layers.

5. Pour 1 to 2 tablespoons of simple syrup over each cake. You can also wrap them in wax paper and secure with a towel or a string tied tight. This will help ensure that the cakes keep their shape while cooling.

6. Freeze the cakes for at least 6 hours, preferably overnight.

Making the Surprise

7. Using a cardboard cake round or a sheet of paper, trace out 3 evenly spaced concentric circles. I measured in 1 inch from the exterior of my cardboard cake round, then in 1 inch more for the second circle, then in 1 inch more for the third.

8. Cut around the line of the outside circle.

9. Place the template on the white cake and use a very sharp knife to cut around it as closely as you can, all the way through the cake. It's very important to get a straight up-and-down cut.

10. Then repeat on the red velvet cake.

11. Cut off the next outside ring from the cardboard template. Center the template on the white cake and cut around the template, again keeping your knife moving perfectly up and down. Repeat on the red velvet cake.

12. Cut off one more ring from the cardboard template and center it on the white cake. Cut around the smallest circle, keeping your knife

making perfectly up and down. Repeat on the red velvet cake. If your cakes feel soft or crumbly, return them to the freezer until firm, as it's important that the cakes are very chilled for the next steps.

13. Cut through the 3 outer circles on each cake with a knife. Do not cut into the center circle. Gently separate the layers into individual sections.

14. Choose a center and place it on a cake stand. Then take the next larger layer of the opposite color cake and place it around the center. Keep alternating the layers until you've assembled a complete cake.

15. Assemble a second cake using the other center and layers.

16. If you're not frosting the cakes immediately, place a piece of parchment paper or wax paper around the perimeter of the cake, then wrap a string or a strip of fabric or ribbon around the paper, and tie snugly to secure the cake (don't squeeze it). Refrigerate until you're ready to frost.

17. Prepare 1 recipe of the buttercream.

18. After the cakes are firm, remove the paper and tie, and cover the cakes in a crumb coat (see page 13).

19. Place in the fridge for 1 hour or in the freezer for 15 minutes.

Frosting and Decorating

20. Using a small offset spatula, cover the cakes in a smooth layer of buttercream.

21. Tint the remaining buttercream blue. Prepare the second batch of buttercream and tint 1 cup of it red. Put frosting into disposable plastic pastry bags. You will fill 2 to 3 bags.

22. Insert a coupler set into a disposable plastic pastry bag and attach a #104 tip. Fill up 2 separate bags with the blue buttercream and place one frosting-filled bag into the bag with the coupler set.

23. Place one of the cakes on a rotating cake stand. Pipe the blue ruffled frosting tops. Hold your tip parallel to the cake at the outside edge with one hand, and begin to apply gentle pressure while slowly turning the cake stand with your other hand. Go around the entire cake, starting the next row of ruffles by overlapping the frosting.

Continue to move in circles until you reach the center of the cake.

24. Center the donkey cutter between the top and bottom of the side of the cake and gently press in to create a guideline. I used a small dog in place of a donkey; I haven't yet run into a donkey cutter. Press the elephant cutter in a couple of inches away. Continue around the cake.

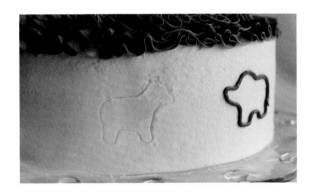

25. Insert a #4 tip into a disposable pastry bag and fill it with the red buttercream. Pipe the outlines of the donkeys and elephants. (I simply elongated the dog's ears into a point so they look more like donkey ears.) Then fill in the donkeys and elephants with red piping.

26. Chill the cakes until serving.

rudolph cake

My niece Inga Grace is supersmart. She's a human sponge, just absorbing information all around her at all times.

I remember when she was three years old she sat me down and recited every one of Santa's reindeers. Every. Single. One. She obviously sensed my awe at her abilities, and I believe it dawned on her that she could do something that I could not do.

She leaned in close, patted my hand, and said, "It's okay, Aunt Amanda. I will teach you."

To this day she teaches me something new every time I see her. Clearly she inherited her smarts from her aunt Amanda.

2 recipes White Cake, page 19

Red, brown, and black gel food coloring; white gel food coloring (optional)

1 tablespoon cocoa powder

1 recipe Basic Buttercream, page 24

¼ cup milk chocolate chips

SPECIAL EQUIPMENT: *4 6-inch round cake pans; 3-inch round cookie cutter; offset spatula*

DIFFICULTY: *Challenging*

Baking

1. Prepare one of the white cake recipes in two 6-inch round cake pans (you may need a little extra baking time when using 6-inch pans).

2. Prepare the batter for the second white cake and divide it in half. Tint half the batter red. (If you add 1 tablespoon cocoa to the white cake batter, you'll be able to use less red food coloring.) Bake 1 white and 1 red layer in two 6-inch round cake pans.

3. Cool the layers to room temperature, then freeze them for at least 6 hours, or overnight.

4. Prepare the buttercream. Place 2 cups of the buttercream in a bowl and color it medium brown with food coloring.

Making the Surprise

5. Place the three white layers on a piece of parchment paper. Make sure the layers are level (see page 9) and all the same height.

6. Center a 3-inch round cookie cutter on top of one of the white layers and press in to create a guideline. Repeat on a second layer.

7. Use a teaspoon to dig out a channel in the first layer. Follow the guideline and cut in about 1 inch deep. (To ensure a consistent depth, I wrap a piece of tape around the spoon. When I insert the spoon into the cake, I know not to go beyond the line of the tape.) Save all the cake scraps in a bowl.

8. Dig a shallower channel, about ½ inch deep, around the guideline on the other layer. This will be the upper part of Rudolph's eyes. Set this layer aside.

9. Place the uncut white layer on a cake stand. Use the 3-inch cookie cutter to press a guideline in the center.

10. Now go back to the layer with the deeper channel and flip it over. Cut a guideline in this layer as well.

11. Gently scoop out cake within the guideline on the layer on the cake stand. You're carving out a bowl in the cake, but keep it shallow—1 to 1½ inches deep, tops. Repeat with the second layer, being careful, as a channel is already cut into this layer on the other side. You're creating Rudolph's nose! Remember to save those cake scraps.

12. Crumble the red cake, removing any hard bits or edges, into a bowl and add 1 to 2 tablespoons of buttercream (either white or red will work). Create a pliable cake mixture (see page 11).

13. Fill the bowl in the layer on the cake stand with red cake mixture. Using an offset spatula, spread a thin layer of white buttercream around the top edge of the layer, taking care not to touch the red cake mixture. Then fill the bowl in the other layer with cake mixture as well.

14. Divide the white cake scraps between 2 bowls. To the first bowl add about 1 teaspoon of white buttercream and a few drops of black gel food coloring. Combine to make a black cake mixture.

15. To the second bowl add about 1 teaspoon buttercream and a bit of white food coloring if you have it on hand. Combine to make a white cake mixture.

16. Roll the black cake mixture into a snake. Carefully flip over the layer on the parchment paper that is filled with red cake mixture. Fill the channel with the black cake mixture.

17. Roll the white cake mixture into a thicker snake and place it on top of the black snake. Gently press the white down around the edges of the black snake.

18. Spread a thin layer of white buttercream around the top edge and center of the layer, avoiding the snakes. Turn over the cake layer with the shallow channel and nestle it onto the snakes so that they fit inside the channel.

22. Place the chocolate chips in a disposable plastic pastry bag. Microwave on high in 30-second bursts until they're fully melted.

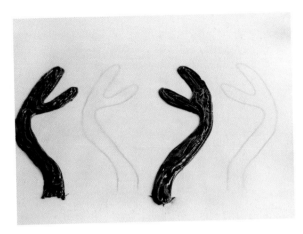

19. Then gently flip the two layers over onto the layer on the cake stand so that the red nose meets in the middle.

20. Cover the cake in a crumb coat (see page 13).

Frosting and Decorating

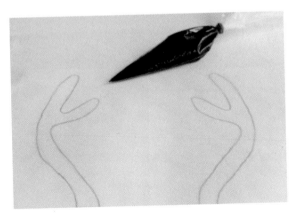

21. Draw an antler shape on paper. Add 2 inches to the bottom of the antlers, since you'll need extra to insert into the cake. Hold the drawing next to the cake and make sure the antlers are the right size. Place the antler tracing underneath a piece of parchment or wax paper on a baking sheet.

23. Cut a very small tip from the pastry bag and pipe the chocolate onto the parchment paper following the lines of the antler drawing. I made an extra set of antlers just in case.

24. Place the baking sheet in the freezer for 1 or 2 minutes or in the fridge for 15 to 30 minutes.

25. Cover the cake in a slightly rough coat of brown buttercream (see page 12).

26. Chill the cake. Right before serving, peel the parchment away from the chocolate antlers and insert them into the cake.

bunny cake

This is one of my favorite cakes ever. I want to hold it close and never let it go, but I might look silly with big pink and white melted chocolate ears stuck on my shirt.

But in all honesty, I wish I could give this cake to you, readers of this book and my blog. As I write and bake cakes and try to maintain my blog, there's one thing I rely on: you. Your comments and support and kindness are more valuable to me than you'll ever know, and I just wish I could give you more than one little cake!

For all of those faithful readers of my blog, I thank you. I thank you with big fluffy frosting bunny tails and I thank you with sky-high pink-and-white chocolate bunny ears.

So, if you decide to make this cake, know it's an extension of my heart. (It would also be supercute for Easter!)

2 recipes White Cake, page 19

Pink, black, and white gel food coloring

1 recipe Basic Buttercream, page 24

White and pink candy melts

SPECIAL EQUIPMENT:
4 6-inch cake pans; cake leveler or long serrated knife; 3-inch and 4-inch cookie cutters; paring knife; offset spatula; disposable plastic pastry bags; #74 decorating tip

DIFFICULTY: *Challenging*

Baking

1. Bake 1 recipe white cake in two 6-inch pans. (You may need to extend the recommended baking time when using 6-inch pans.) Cool to room temperature.

2. Level the cakes (see page 9) and freeze for about 1 hour.

3. Prepare the batter for a second white cake. Divide the batter between 2 bowls. Tint the batter in one bowl pink.

4. Bake the cake in two 6-inch pans and cool to room temperature. Level the cakes to match the others and freeze for about 1 hour.

7. Remove the white cake cone and set it in a scrap bowl—you'll be using it for the bunny's eyes.

5. Place one white layer on a cake stand, off center. Place the pink layer next to it on a sheet of parchment paper. Then center a 3-inch round cookie cutter in on the white layer and press in slightly to make a guideline. Repeat on the pink layer.

8. Cut out a cone the same size in the pink layer and place the pink cone into the white layer.

6. Insert a sharp paring knife at about a 45-degree angle on the guideline on the white cake. Aim for the center, then slowly pull the knife around the cake, following the guideline.

9. Place the other 2 white layers on a sheet of parchment paper. Center a 4-inch round cookie cutter on each layer and press in slightly to make a guideline.

10. On one layer, dig out a channel about ¼ inch deep on the guideline (see the cake on the left). On the other layer, dig out a channel about ½ to ¾ inch deep. Add the crumbs to the cake scrap bowl.

11. Crumble the rest of the pink cake into a bowl, removing any dark or hard spots.

12. Add about 1 tablespoon buttercream and ¼ teaspoon black gel food coloring. Mix until you have a cake mixture of your desired color. You can add more frosting or food coloring as needed.

13. Add about 1 teaspoon of buttercream to the white cake scrap bowl. (I also added a bit of white gel food coloring to make the color bright.) Mix until fully combined.

14. Roll the black cake mixture into snakes thick enough to fill the deeper cavity in the white layer.

15. Lay them in and connect them to fit evenly.

16. Roll out a smaller white snake from the white cake mixture and flatten it a little with your finger, but make sure it's still at least ¼ inch thick.

18. Using an offset spatula, spread a thin layer of buttercream around the white edge of the layer on the cake stand. Be careful to avoid the section of pink cake.

17. Place the white cake mixture on top of the black snake, covering it completely.

19. Set the layer with the black snake on top. Cover the top of the cake with a thin layer of buttercream.

20. Turn over the remaining cake layer and lay it on top of the cake. Make sure the snake fits snugly into the carved-out top layer.

21. Cover the cake in a crumb coat (see page 13) and chill for at least 1 hour before decorating.

Frosting and Decorating

22. Insert a #74 decorating tip into a disposable pastry bag and fill it with buttercream.

23. Applying light pressure, make small, short bursts with the pastry bag. They can be irregular in shape and size—just try to keep them about the same thickness. Do this over the whole cake.

24. To make the bunny's tail, just pipe out a big ball of frosting. Tint ½ cup of the buttercream pink and place it in a pastry bag with a #74 tip. Pipe out the same pattern as that on the cake to cover the tail.

25. Now it's time to make the bunny ears! Place 1 cup white candy melts in a clear plastic pastry bag and microwave in 30-second increments until the chocolate is mostly melted. Massage the bag with your hands until it's fully melted.

26. Place a sheet of parchment paper or wax paper on a baking sheet. If you like, you can draw out your design on a piece of paper and slip it behind the parchment paper. (See the reindeer antlers on page 204 as an example.)

27. Cut a very small tip off the bag. Pipe out a white inner ear, making sure to extend the bottom of the ear so that you'll be able to insert it into the cake. Repeat to make the second ear. (I made 4 ears just in case there were any breaks.)

28. Melt 1 to 2 cups pink candy melts and repeat the process to create the outside of the ears.

29. Refrigerate the ears on the baking sheet for 30 minutes or freeze them for 2 minutes.

30. Chill the cake. Just before serving, peel the parchment away from the ears and insert them into the cake, one at left center and one at right center, facing away from the tail. The ears should stay in place while serving, but they usually don't last that long. I know my kids are always excited to see who will get them! The extra ears are handy, too, in case of sibling rivalry.

candy cane cake

Truth be told, I'm not a fan of candy canes. I almost feel as if they should be used for decorative purposes only: hanging on trees, perched on the edge of a stocking, taped to cards from aunts, uncles, and grandmas. But the whole process of eating one frustrates me. I like the flavor, so I want to bite it, but then it gets all sharp and stuck in my teeth, which isn't fun. And simply licking is far too slow.

This cake is my happy compromise. Flavor? Check. Design? Check. Ease of gobbling an entire piece as quickly as I want? You betcha.

2 recipes White Cake,
page 19

1 recipe Red Velvet Cake,
page 21

1 recipe Peppermint
Buttercream (use
peppermint extract in place
of vanilla extract in the Basic
Buttercream recipe,
page 24)

Red gel food coloring

SPECIAL EQUIPMENT:
*6 8-inch round cake pans;
long serrated knife; offset
spatula; long skewer (optional);
disposable plastic pastry bags;
2D decorating tip*

DIFFICULTY: *Challenging*

Baking

1. Bake 4 layers of white cake and 2 layers of red velvet cake in 8-inch round cake pans. (There will be 1 white layer left over; wrap it in plastic, cover in foil, and freeze it for up to 3 months.)

2. Cool the layers to room temperature, then freeze for at least 3 hours. The cake should be firm but not crumbly.

3. Prepare the buttercream. Tint 1 cup of the buttercream red to match the red velvet cake.

Making the Surprise

4. Place 1 white layer on a sheet of parchment paper. Using a long serrated knife and starting at the upper-right-hand corner, cut from one edge to the lower left-hand corner at a 45-degree angle.

5. Next, take the top cut white layer and carefully flip it over. Set it on top of the bottom layer so that the thickest parts are on top of each other, creating a steeper-angled layer. Once you get the right snug fit, remove the top, spread a thin layer of coordinating buttercream on the bottom layer, and replace the top. This will hold as a glue and keep cakes firmly in place.

6. Repeat steps 4 and 5 to create the top of the cake, and set this steep cake aside.

8. Center a red velvet layer on top of the white cake. It will seem off center because the sides will not perfectly match up, but it's the centers of the cake that need to align. Spread a thin layer of buttercream on top.

9. Repeat, alternating a white cake layer followed by the second red velvet layer.

7. Transfer the bottom steep layer to a cake stand or a cake plate and, using an offset spatula, spread a thin layer of buttercream on top.

10. Flip the top steep layer and lay it on the cake so that the top of the cake is now level. Voilà!

11. If your cake seems at all unsteady, you can insert a long thin skewer into the cake for added stability. You can also freeze the cake for at least 1 hour or up to overnight.

12. When the cake is steady, spread on a crumb coat (see page 13), filling the gaps between the layers to create smooth sides.

13. Chill the cake in the fridge for at least 1 hour to set the crumb coat.

Frosting and Decorating

14. Place buttercream into a plastic disposable pastry bag fitted with a 2D tip.

15. Apply slight pressure to the bag of buttercream and pipe a single star on the cake. Repeat over the whole cake.

CELEBRATING LIFE

you may have noticed this already, but I'm slightly quirky. I tend to speak faster than I think and think slower than I speak.

Celebrating life is a true representation of how random my brain is. One day I'm dreaming of vampire-themed cakes and the next I'm pondering how to incorporate my kids' fashion into my sweet treats.

vampire cake

One of the greatest aspects of putting this book together was the people I have met, and one of those very special people is the photographer, Susan Powers. Her blog, *Rawmazing.com,* is the most beautiful compilation of her work! Not to mention healthy. And after making all the cakes for this book, we *needed* healthy!

Her talent was especially obvious when I saw her shoot this cake. She captured the mood that I envisioned perfectly! Because of her amazing skills, I think this is one of my favorite cakes in the book. You can make this delicious cake and sit down to watch your favorite vampire-themed show!

1 box dark chocolate cake mix (or add dark brown food coloring to the recipe for Chocolate Cake, page 20)

½ recipe White Cake, page 19

½ recipe Chocolate Buttercream, page 25

Dark brown gel food coloring

½ recipe Basic Buttercream, page 24

Red Glaze

SPECIAL EQUIPMENT:
2 8-inch round pans; cupcake tin (optional); offset spatula; 4½-inch round cookie cutter; paring knife

DIFFICULTY: *Easy*

red glaze

2 cups powdered sugar

1 tablespoon corn syrup

1 to 2 tablespoons red gel food coloring

Up to ¼ cup milk, for thinning

Combine the powdered sugar and corn syrup and add red gel food coloring to get the shade you want. If necessary, add milk to create a somewhat runny consistency—think runny pancake batter.

Baking

1. Bake the dark chocolate cake in two 8-inch round cake pans. Cool to room temperature, then chill for about 6 hours (unlike some other surprise-inside cakes, this one needs no more than 6 hours—this is a very simple design).

2. Bake the white cake as cupcakes or in whatever pan you like. You'll only need about ¼ of the cake.

3. Make the chocolate buttercream and add dark brown food coloring to make it the same color as the chocolate cake.

Making the Surprise

4. Place one chocolate layer on a cake stand. Cover in a layer of chocolate buttercream and top with the second chocolate layer.

5. Gently insert a 4½-inch cookie cutter into the center of the cake to create a guideline.

6. Use a small, sharp knife to create a V-shaped valley in the cake. Working ½ inch from the guideline on the inside, insert the knife at an

angle toward the guideline and cut carefully around the entire cake, going no deeper than the first layer. Next, insert the knife ½ inch outside the guideline at a 45-degree angle toward the other cut to make a V, then cut around the cake.

7. Remove the cake between the cuts to make a channel.

8. Crumble a quarter of the white cake into a bowl, removing any dark or hard bits. Combine 1 cup of the crumbled white cake and 1 to 2 teaspoons of white buttercream in a small bowl and combine to make a cake mixture (see page 11). Make a snake out of the mixture, cut it into sections for ease of use, and flatten two sides into a point that will fit the channel you created in the chocolate cake. This will be your pointed vampire tooth.

9. Fill the cavity (ha-ha!) with sections of the cake mixture.

10. Cover the top of the cake in the chocolate buttercream. Freeze the cake for at least 15 minutes to help the buttercream set.

Frosting and Decorating

11. Immediately before serving, pour the blood (red glaze) onto the center of the cake and let it spill over the edges. Better yet, do this in front of the guests . . . it's truly a sight to behold!

vampire cupcakes

Want to make a fun treat? Make vampire cupcakes! Simply level out a chocolate cupcake, then create a tooth channel in each the same way you would for a full-size cake. Fill the channel with white frosting, then cover in dark chocolate frosting. Chill before you serve. Be sure to let the recipient know there's a surprise inside!

argyle cake

When I was growing up, the buzzword was *preppy*. I wanted so badly to be preppy! And to me, nothing screamed prep more than a classic argyle sweater with matching socks. Yes, *matching* socks. Although I haven't matched my sweater to my socks in decades, I've never lost my fondness for that particular trend. And even though it might not be cool to be preppy anymore, I can still dress my kids up in argyle sweaters and matching socks! At least for a few more years, or until they start to boycott Mom's choice in favor of what's actually in style.

This recipe makes two 2-layer cakes.

1 recipe White Cake,
page 19

1 recipe Chocolate Cake,
page 20

1 recipe Basic Buttercream,
page 24

1 recipe Chocolate
Buttercream, page 25
(optional)

Turquoise or pink and
brown gel food coloring

SPECIAL EQUIPMENT:
*4 6-inch round cake pans; cake
leveler or long serrated knife;
rotating cake stand (optional);
3-inch round cookie cutter;
3½-inch round cookie cutter;
offset spatula; disposable
plastic pastry bags; #2 and
#4 decorating tips*

DIFFICULTY: *Medium*

Baking

1. Bake 2 layers of white cake and 2 layers of chocolate cake using the 6-inch cake pans (you may need to extend the recommended baking time when using 6-inch pans). Cool the cakes to room temperature.

2. Level all 4 layers to the same height (see page 9).

3. Refrigerate the layers for 6 hours or freeze them for 1 hour.

Making the Surprise

4. Place 1 white layer on a piece of parchment on a flat surface or rotating cake stand. Center a 3-inch round cookie cutter on the layer and press slightly to create a guideline. Repeat with the other 3 layers.

5. Insert a toothpick or skinny skewer into the dead center of the cake on the stand.

6. Insert a very sharp paring knife into this layer at the guideline at a 45-degree angle. Cut all the way to the skewer. With the knife still in cake, place one hand flat on the cake and slowly cut around, following the guideline. Move as slowly as you need, and be as precise as you can. If the cake feels too soft or crumbly, chill it again to make it easier to work with. Thirty minutes in the freezer should do the trick!

7. Repeat to cut matching cones out of the other 3 layers.

9. Freeze all the layers for at least 1 hour.

10. Prepare the basic buttercream and tint 2 cups of it turquoise (or pink, if you prefer that color scheme).

11. Remove the cakes from the freezer and place the layers on parchment paper. Carefully turn the layers over so that the uncut sides are facing up. Then center a 3½-inch round cookie cutter on one layer and press in slightly to create a guideline. Insert a toothpick or skinny skewer into the dead center of the layer.

8. Remove the cones and place the white cones into the chocolate layers and vice versa.

12. Insert a very sharp paring knife into the layer at a 45-degree angle. Cut all the way to the skewer. With one hand flat on the cake, slowly cut around the guideline. Repeat with the other 3 layers.

13. Carefully remove the newly cut section and, using an offset spatula, frost the interior with a thin layer of turquoise buttercream. Replace the cone of cake. Repeat with the remaining 3 layers.

14. Place one white layer on a cake stand with the chocolate center side up. Spread a layer of white buttercream around the top perimeter of the layer, not touching the chocolate center. Carefully set the other white layer on top of the first layer.

15. Cover the cake in a chocolate crumb coat (see page 13).

16. Create a second 2-layer cake with the chocolate layers, using ¼ cup of chocolate buttercream around the perimeter of the bottom layer.

17. Crumb-coat the cake with white buttercream.

Frosting and Decorating

18. Cover the white cake in a smooth layer of chocolate buttercream, and the chocolate cake in a smooth layer of white buttercream (see page 12).

19. Insert a #4 decorating tip into a disposable pastry bag. Fill a second bag with white buttercream and a third bag with chocolate buttercream. Cut the tip off the bag of white buttercream and slip it into the pastry bag with the #4 tip. Pipe out a row of diamonds around the base of each cake. Pipe out white diamonds on the brown-frosted cake. Replace the bag of white buttercream with the chocolate buttercream and pipe out a bit of frosting until the tip is clear of any white frosting. Pipe out brown diamonds on the white-frosted cake. Insert a #2 tip into a pastry bag and fill it with the turquoise (or pink) buttercream. Pipe Vs and inverted Vs above and below where the diamonds intersect.

20. Chill the cakes before serving.

arrow cakes

I have a confession. I originally wanted to call these Naughty and Nice cakes, but it was just a little too suggestive for me. I mean, was I going to serve it with a side of whipped cream and cherries and put a parental advisory on the page?

Next I considered calling them Heaven and Hell cakes, but I didn't want to make a guest say, "I'll take a slice of Hell, please. Can you make sure I get a slice with horns?"

When you make these cakes—and you really should—you're welcome to call them whatever you like!

This recipe makes two 2-layer cakes.

Baking

1. Bake 4 white layers and 4 chocolate layers in 8-inch round cake pans. Chill for at least 6 hours, preferably overnight.

2. Prepare the buttercreams.

3. Place one white layer on parchment paper. Using an offset spatula, spread about ¼ cup of white buttercream on the top of the cake. Place another white layer on top. Repeat to make a second 2-layer white cake. Chill for at least 15 minutes.

4. Repeat with the chocolate layers using chocolate buttercream to create two 2-layer cakes. Chill for at least 15 minutes.

Making the Surprise

5. Starting with a white cake, center and gently press a 4½-inch round cookie cutter just barely into the cake and then remove. This guideline will let you know where to insert your knife. Insert a skewer into the center of your cake and press all the way to the bottom. Insert a sharp knife into your guideline at about a 45-degree angle until it hits the skewer. Slowly and steadily cut around the whole cake, keeping the knife on the guideline and the tip of the knife touching the skewer.

¼ cup white chocolate chips

Gold sprinkles

¼ cup chocolate chips

SPECIAL EQUIPMENT:
8 8-inch round cake pans; offset spatula; 4½-inch round cookie cutter; skewer; paring knife; 2-inch round cookie cutter; cake leveler; disposable plastic pastry bags

DIFFICULTY: *Medium*

6. Carefully remove the cone you just cut—this will be the point of the arrow in the chocolate cake. If you find that the cake is getting soft or crumbly, put it back in the freezer before removing any pieces.

7. Repeat this process on one of the chocolate layer cakes.

8. Gently place the white cake cone into the chocolate cake cavity and the chocolate cake cone into the white cake cavity.

9. Center a 2-inch round cookie cutter on top of the second white layer cake and press it in slightly to make a guideline. Using a sharp knife, cut a column through the entire cake following the guideline. Remove the column very carefully, as you will be inserting it into the chocolate cake. Freeze the cake as needed if it starts to get crumbly.

10. Repeat with the second chocolate cake. Carefully place the chocolate cake column into the white cake and the white cake column into the chocolate cake.

11. To make the upward-pointing arrow cake, simply place one of the 2-layer cakes that has the column removed on a cake stand. Cover the top in coordinating buttercream—chocolate buttercream on the chocolate cake portions, and white buttercream on the white cake.

12. Making sure the point (the narrower part of the cone) is facing up, place the coordinating 2-layer cake on top. Cover in a crumb coat (see page 13).

13. To make the downward-pointing arrow cake, place remaining layer cake with the cone removed on a cake stand with the point (the narrower part of the cone) on the bottom. Cover the top of the cake with coordinating buttercream. Place the coordinating 2-layer cake on top. Cover in a crumb coat.

Frosting and Decorating

14. After the crumb coat has set, spread a smooth layer of matching buttercream on the cakes.

15. To make the halo: Using an 8-inch round cake pan, trace out a circle on parchment or wax paper. Flip the paper over when you're ready to pipe so you don't get any ink on the chocolate and lay it flat on a baking sheet for stability.

16. Put the white chocolate chips into a plastic disposable pastry bag and heat in the microwave in 30-second increments until melted. Cut a very

small tip off of the bag and pipe the chocolate into a solid ½-inch circle (you'll have to go around several times). Immediately cover the white chocolate with gold sprinkles.

17. Place the baking sheet in the freezer for 2 minutes or in the fridge for 15 minutes. Peel the paper away from the chocolate and gently place the halo over the edge of the cake.

18. To make the horns: Draw basic horn shapes onto a piece of parchment or wax paper; make them a little longer than you'd want them because the flat end will be inserted into the cake. I make 2 left-facing horns and 2 right-facing horns just in case one breaks. Flip the parchment paper over so that you don't get any ink on the chocolate and lay it flat on a baking sheet for stability. (Take a look at the antlers on the Rudolph Cake on page 204 to see this process.)

19. Put the chocolate chips into a disposable plastic pastry bag and heat in the microwave in 30-second increments until melted. Cut a very small tip off the bag and fill in the outline of your drawing with the chocolate, going back and forth so that the horns are stable and solid.

20. Place the baking sheet in the freezer for 2 minutes or in the fridge for 15 minutes.

21. Gently peel the paper away from the horns and insert them into the cake.

22. Serve immediately.

football cake

My folks grew up in Bemidji, Minnesota; we still have lots of extended family there and visit as often as possible. I have great memories of going to my grandma and grandpa's house as a child, and I especially loved to hear about what it was like growing up where the Minnesota Vikings had training camp. See, my grandpa owns a gas station on the main road in Bemidji, and he often had the pleasure of meeting players, coaches— pretty much everyone associated with the team. (Of course this was back in the day when all gas stations were full-service.) Some of the players even came out to their house for dinner!

So as you can see, I was raised to love the game of football. And while my active interests have changed to things like babies and cake, I still enjoy watching a game with my husband when the Minnesota Vikings are on.

Feel free to incorporate your team's colors into the cake or frosting!

Baking

1. Prepare the white cake batter, adding 2 to 4 drops of yellow food coloring to the batter (or prepare a yellow cake recipe instead). Bake two 6-inch round cake layers (you may need to extend the recommended baking time when using 6-inch pans). Cool to room temperature. If your cakes are not level, level them now (see page 9).

2. Freeze the layers for at least 6 hours, preferably overnight, to make the cake texture firm, not crumbly.

3. Bake the chocolate cake in two 6-inch layers. Cool to room temperature. You'll need only one layer; you can wrap the other in plastic wrap and foil and freeze it for up to 3 months.

4. Prepare the buttercream.

1 recipe White Cake, page 19 (or your favorite yellow cake recipe)

1 recipe Chocolate Cake, page 20

1 recipe Basic Buttercream, page 24

Yellow, white, and brown gel food coloring

SPECIAL EQUIPMENT:
4 6-inch round cake pans; cake leveler or long serrated knife; 3-inch round cookie cutter; paring knife; skewer (optional); rolling pin; offset spatula

DIFFICULTY: *Medium*

Making the Surprise

5. Press a 3-inch round cookie cutter into the center of one of the layers to create a shallow guideline.

6. Using a sharp paring knife, go in and add 2 points to the circle shape to extend the circle into a football-shaped guideline. Use a skewer to help keep the cuts even if you like. Repeat on the second layer.

7. Carve out the shape of a football. Make the deepest cut in the cake at the center point, and graduate up and out. Reserve the cake scraps in a bowl.

8. Repeat to cut out a football shape on the second layer.

9. To make the football, crumble one chocolate cake layer and remove the hard bits. Mix in 1 or 2 tablespoons of buttercream to create a pliable cake mixture (see page 11).

10. Add 1 or 2 tablespoons buttercream and 1 to 4 drops white food coloring to the bowl of reserved yellow cake scraps and combine to make the cake mixture. Roll the cake mixture thin on a piece of parchment.

12. Lay out the seam pattern in one of the cake layers to make sure it fits properly. Cut and adjust as needed, then remove it from the cake and set it aside.

11. Cut the cake mixture into a single 1-inch strip and five ¼-inch strips.

13. Form a rough football shape out of chocolate cake mixture to fit the cavity in the cake. Adjust as needed; the cake mixture should be level with the top of the layer.

14. Remove the football half, turn it over, and make 5 short cuts into the top to fit the 5 white cross-seams. Nestle the 5 shorter white cake strips into the top of the football. They should fit firmly inside the chocolate cake football. Then make one

long cut lengthwise across the 5 shorter strips to fit the longer, wider cake strip. Insert the longer white strip into this cut.

15. Turn the football half over and nestle it into the cavity of the cake layer, white strips down. This will be the top layer of the cake.

16. Fill both layers of yellow cake completely with the chocolate cake mixture. Place a dab of buttercream on the bottom layer so you know which is which.

17. Tint a few tablespoons of buttercream yellow (to match the cake interior) and spread it around the perimeter of the bottom cake layer using an offset spatula. Be careful to not get frosting on the chocolate football halves.

18. Gently invert the top layer and place it over the bottom layer, taking care to match the ends of the football. You'll want to cut the first slice from the side of the football to show the shape, so keep track of the front. Cover the cake in a crumb coat (see page 13) and place it on a cake stand.

Frosting and Decorating

19. Cover the cake in a smooth layer of buttercream (see page 12). Tint the remaining buttercream brown.

20. Insert a #2 decorating tip into a disposable pastry bag and fill it with the brown buttercream. Pipe outlines of people cheering around the bottom perimeter of the cake. If you like, you can draw this design on paper first. Place the drawing behind a piece of wax paper and practice piping. You can also use a toothpick to draw the design directly on the cake. If you make a mistake, simply smooth out the frosting and start over.

21. Fill in the outlines with chocolate buttercream.

22. Chill the cake until serving.

sailboat cake

Living in Minnesota, land of 10,000 lakes, is a great asset when you're a water lover. My husband is an avid fisherman and even makes his own custom fishing rods! His passion for fishing and water has certainly carried on to our children. We all really like swimming and, fishing and, of course, boats. One of my favorite feelings is the spray of water when cruising around a lake in the hot summer. This sweet little sailboat cake is designed for anyone who loves to dream of playing in the water and long days spent fishing with Dad. It would also be perfect for a little boy's birthday!

Baking

1. Prepare the first white cake recipe and tint the batter a light blue. Bake the cake in two 8-inch round cake pans. Set aside to cool, then level the cakes if necessary (see page 9). Freeze the layers for at least 6 hours.

2. Prepare the second white cake recipe. Divide the batter equally between 2 bowls and color one bowl red. Bake the white and red layers in 8 × 8-inch square pans or 8-inch round pans. Cool the layers to room temperature and wrap them in plastic wrap until you're ready to use them.

Making the Surprise

3. Place a 4½-inch round cookie cutter in the center of each blue layer and gently press in to create a guideline.

2 recipes White Cake, page 19

1 recipe Basic Buttercream, page 24

Blue and red gel food coloring

Blue or turquoise sprinkles

SPECIAL EQUIPMENT: *2 8-inch round cake pans; cake leveler or long serrated knife; 4½-inch round cookie cutter; tugboat or duck cookie cutter; roller; offset spatula; disposable plastic pastry bags; #4 decorating tip*

DIFFICULTY: *Medium*

4. Using a tugboat or duck cookie cutter, gently insert it at an angle into one of the layers, centering the cookie cutter on the guideline.

5. Insert the cutter only halfway—don't go too deep. Slowly and carefully pull the cookie cutter around the cake to cut a shallow well, keeping it centered and at a uniform depth.

6. Remove the cut cake and reserve it in a bowl.

7. Working on the second blue layer, insert a small, sharp knife about 1 inch inside the guideline. You're creating a V or valley in the cake, so angle the cut toward the outside of the cake. Cut around the entire cake.

8. Make a cut ½ inch outside the guideline, this time cutting toward the center to finish the V cut.

9. Remove the excess cake from the channel and place it in the bowl with the reserved blue cake scraps.

10. Prepare the buttercream.

11. Crumble the white layer into a large bowl, removing hard bits and dark pieces. Add 2 to 4 tablespoons of buttercream and combine to make a pliable cake mixture (see page 11).

12. Make a second batch of cake mixture using the red cake.

13. Carefully press the red cake mixture into the hollow made with the tugboat cookie cutter. Be sure to fill the whole cavity evenly.

14. Roll the white mixture into a small snake. Press it gently on a piece of parchment so that one side is flatter. Pinch the top into a point to create a triangle that will fit in the valley shape of the other blue layer.

15. Place the white cake mixture in the valley. Continue until the channel is filled.

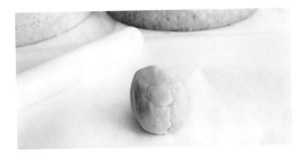

16. Add about 1 teaspoon of buttercream to the bowl of reserved blue cake scraps. Mix to create a pliable cake mixture.

17. Place the cake mixture on one side of a piece of parchment and fold the parchment over the top. Roll it out to make a sheet about ¼ inch thick.

18. Cutting the sheet of cake mixture to fit, lay pieces over the red cake mixture to cover it thoroughly. The outlines can be rough.

19. Mix a little blue food coloring into ½ cup buttercream to match the shade of the cake and spread it evenly on top of the blue/white cake layer (at left). This is the top of the boat.

20. Transfer the blue/red layer to a cake stand. Gently turn over the blue/white layer and set it on top of the blue/red layer.

21. Cover the cake in a crumb coat (see page 13).

22. Chill the cake for at least 1 hour in the fridge or 15 minutes in the freezer.

Frosting and Decorating

23. Cover the cake in a smooth coat of white buttercream (see page 12). Insert a #4 decorating tip into a disposable pastry bag and fill it with white buttercream. Pipe waves around the side of the cake. Cover the cake with blue sprinkles, letting them fall over the sides of the cake onto the cake stand.

24. Chill the cake until serving.

smiley face cake

Hi, my name is Amanda and I have an addiction to emoticons. ☺ I'm afraid that at this point I'm unable to convey my true emotional state without them.

I talk often with my friend Jessica Walton about Twitter, Facebook, Instagram—all the social outlets I enjoy using. Although she and I text and e-mail each other, I sometimes share other social media shortcuts with her. I add #hashtags to my thoughts and totally overuse emoticons. Every sentence ends in a smiley face. Every thought is followed by a sometimes inappropriately placed hashtag.

#icanthelpmyself

But she cleverly rolls with it. And here's a true testament to our friend-ship . . . Jessica doesn't really like sweets. Cake is probably at the bottom of her list of favorite things to eat. It's a good thing she can pound out a witty hashtag.

Baking

1. Bake 1 recipe of the white cake in two 6-inch pans. (You may need to extend the recommended baking time when using 6-inch pans.) Cool the cakes to room temperature, then freeze them for at least 6 hours.

2. Prepare the second recipe of white cake and divide it between 2 bowls. Mix 4 to 6 drops of red gel food coloring to one bowl and bake it in the pan of your choice (you can even make cupcakes—you'll only need a couple of them for this recipe, and the others can be a treat). Bake the white batter in a 6-inch round cake pan. Cool the white layer and freeze it for at least 6 hours, preferably overnight.

3. Prepare the buttercream.

4. In a medium bowl, combine 2 red cupcakes or 1 heaping cup of crumbled red cake and 1 teaspoon buttercream. Combine to make a cake mixture (see page 11); you want a Play-Doh-like consistency. Add a bit more buttercream as needed. Store in a plastic bag or airtight container until you're ready to assemble the cake.

2 recipes White Cake, page 19

½ recipe Basic Buttercream, page 24

Red, yellow, blue, and green gel food coloring

SPECIAL EQUIPMENT:
3 6-inch round cake pans; cupcake tin; 2-inch and 2½-inch round cookie cutters; paring knife; roller; small offset spatula; disposable plastic pastry bags; #7 decorating tip

DIFFICULTY: *Medium*

Making the Surprise

5. Center a 2-inch and 2½-inch cookie cutter on top and gently press them to create two guidelines. Carefully flip the layer over and make the same guidelines on the other side. Place one layer on a piece of parchment.

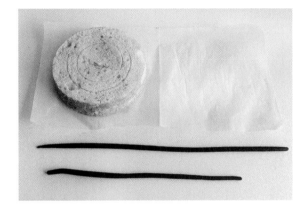

6. Roll out about a quarter of the red cake mixture to make a thin snake about 12 inches long. Make a second snake that's about 14 inches long.

7. Lay the shorter snake in the inner guideline circle and the longer snake in the outer guideline circle. Trim them to fit as needed.

8. Place another layer on top and gently press down on it. (I chose to not use any buttercream between the layers of this cake. If you do so, match the buttercream color to the cake color.)

9. Carefully flip the 2-layer cake. Starting at the center point between the two guidelines, insert a paring knife about ¼ inch and cut out to the exterior guideline. This is a very subtle cut. Go around the entire outer guideline to make a shallow channel. If the cake feels too soft or crumbly, return it to the freezer for at least 30 minutes.

10. Repeat the cut, this time on the inner guideline.

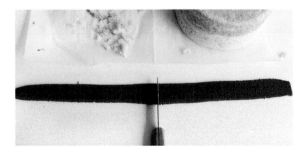

11. Roll out a small amount of red cake mixture to make a flat snake about ¼ inch thick and 14 inches long. Cut it into sections for easier handling.

12. Gently lay a section of the snake in the channel you've just carved. Very gently press back the side of the cake outside the channel, making sure the red cake mixture gets tucked down into the crevice. Do this as well on the inside of the channel, pulling the cake center away from the channel and tucking in the red cake mixture.

13. Repeat until the channel is filled. You've created an upside-down smile!

14. Place the last layer on top of the cake, then gently flip the 3-layer cake over and place it onto a cake stand.

15. Freeze the cake for a bit if it's feeling crumbly. Cover the cake in a crumb coat (see page 13).

Frosting and Decorating

16. Cover the cake in a smooth layer of white buttercream (see page 12).

17. Divide the remaining buttercream among 4 bowls. To each add 1 to 3 drops of red, yellow, blue, or green food coloring.

18. Insert a #7 decorating tip into a disposable pastry bag. Fill 4 other bags with the colored frostings. Select a color, cut the tip off of the pastry bag, and insert it into the pastry bag fitted with the decorating tip. Pipe dots on the cake in the pattern of your choice. (I chose to make a curved line that spanned across the entire cake.) Repeat with the other colors.

19. Let the buttercream set for 15 minutes, then (with a clean finger) press down the pointed tips on the buttercream dots to flatten them.

20. Chill the cake until serving.

tree cake

This sweet little tree cake was created on a cold and dreary winter day in Minnesota. I often go into sensory deprivation after a long winter and will take any bit of green I can get! This cake fit the bill, as I got to *smell* a delicious cake baking, *touch* the velvety cake texture with my hands, *see* a beautiful tree created, and *hear* the delighted exclamations when it was served!

Sensory overload in a good way. The perfect cake for any nature lover—or someone who needs a dose of summertime, fast!

2 recipes White Cake, page 19

1 recipe Chocolate Cake, page 20

1 recipe Basic Buttercream, page 24

Green gel food coloring

½ recipe Chocolate Buttercream, page 25

SPECIAL EQUIPMENT:
4 8-inch round cake pans; cupcake tin (optional); offset spatula; 2-inch and 4½-inch round cookie cutters; disposable plastic pastry bags; coupler; #4 and #2 decorating tips

DIFFICULTY: *Medium*

Baking

1. Bake 4 white cake layers in 8-inch pans.

2. Bake the chocolate cake as cupcakes or in the pan of your choice.

3. Prepare the basic buttercream.

4. Place 1 cake layer on a sheet of parchment paper and spread ½ cup of buttercream over the top using an offset spatula. Add a second layer to make a 2-layer cake. Freeze the cake for at least 6 hours or preferably overnight.

5. Freeze the third layer and place the fourth layer in an airtight container at room temperature.

Making the Surprise

6. Gently press a 4½-inch cookie cutter into the center of the 2-layer cake. You'll be carving out the top of the tree within this guideline.

7. Using a small spoon, carve out cake within the guideline circle, but don't go past it. The general shape you're going for is a dome, but you really can play around with this and make it yours—treetops are not perfectly dome-shaped!

8. Place ¼ cup of buttercream in each of two medium or large bowls. Add enough green food coloring to each bowl to make 1 bowl of light green and 1 bowl of dark green.

9. Divide the room temperature cake layer in half and crumble half into each of the bowls with the frosting, removing the hard bits and dark parts. Use your hands to fully incorporate the frosting into the crumbled cake to make a cake mixture (see page 11).

10. Use the two shades of green cake mixture to make the leaves of the tree. Place random lumps of both cake mixtures into the cutout dome shape, alternating the colors to add visual depth to the tree.

11. Fill the cavity until it's even with the top of the cake.

12. Next, make the tree trunk. Place the remaining frozen cake layer on a sheet of parchment. Press a 2-inch round cookie cutter through the center of the cake, going all the way through. If the cutter doesn't go all the way through, finish the cut with a thin paring knife.

chocolate cake mixture, letting a little bit of cake extend above the hole to give the area where the trunk meets the branches a more natural look.

13. Flip the cake and make small grooves around the edges of the hole to create a more organic base to the tree. Keep the grooves shallow—don't cut more than halfway into the layer.

14. Crumble the chocolate cake into a bowl, removing the hard bits, and add up to ¼ cup buttercream. Use your hands to fully incorporate the frosting into the crumbled cake to make a cake mixture.

16. Spread a layer of buttercream across the top of the white part of the layer, taking care not to get any on the chocolate cake mixture. Then gently flip the 2-layer cake and set it on top. The green leaves will meet the chocolate trunk in the middle.

17. Using a small offset spatula, spread a crumb coat over the cake (see page 13). Chill the cake for at least 1 hour in the refrigerator or 15 minutes in the freezer to set the tree.

15. Fit some chocolate cake mixture into the grooves, as these will be hard to reach once the cake is flipped. Then gently flip the cake layer onto a cake stand. Continue to fill the hole with

Frosting and Decorating

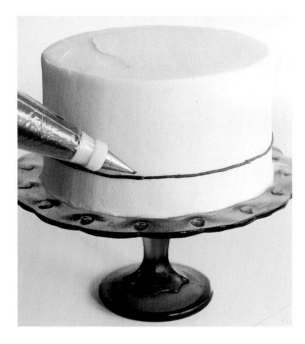

18. Cover the cake in a smooth coat of buttercream (see page 12).

tips and tricks

It's good to clean your cake stand prior to chilling a cake! Then when you go to do the final decorations, you won't have to worry about pesky frosting spots and crumbs.

19. Prepare the chocolate buttercream.

20. Insert a coupler into a disposable pastry bag and attach a #4 decorating tip. Fill the bag with chocolate buttercream. Pipe a line around the cake about a third of the way from the bottom.

21. On a piece of paper, draw out a few different bird shapes. Try doing an Internet image search for "bird on a wire" to get inspiration. Practice piping out the shapes on a sheet of parchment set over the paper, or if you prefer, outline the birds directly on the cake with a toothpick.

22. Pipe the bird shapes onto the cake, outlining them and then filling them in. Change to a #2 decorating tip and add a little beak to each bird.

23. Chill the cake until serving.

bee cake

One of my favorite cookie decorators in the world is Bridget from bakeat350 .blogspot.com. She's just one of those people you love to love. She oozes Southern charm and grace; and, of course, massive amounts of cookie-decorating talent!

Back in 2008, Bridget posted a sweet little bee cookie. Honestly, I wasn't even thinking about cakes in 2008. I didn't even know how to turn on my oven! But for some reason her little bee cookie stuck with me, and when I started making cakes with surprises in them, I knew I wanted to tackle the bee.

This cake is dedicated to Bridget, who could not *bee* any sweeter. Could not *bee* any nicer. Could not *bee* any friendlier. Thanks so much for *bee-ing* an amazing friend, Bridget!

1 recipe Chocolate Cake, page 20

1 recipe White Cake, page 19

1 recipe Honey Buttercream, page 24

Yellow and brown gel food coloring

¼ cup chocolate chips

SPECIAL EQUIPMENT: *2 6-inch round cake pans; cake leveler or long serrated knife; 9 × 13-inch cake pan; 2-inch round cookie cutter; paring knife; offset spatula; disposable plastic pastry bag; decorating tip*

DIFFICULTY: *Challenging*

Baking

1. Bake the chocolate cake in two 6-inch round cake pans (you may need to add a little extra baking time for 6-inch pans). Cool to room temperature, then freeze for at least 6 hours or preferably overnight.

2. Level the cake layers (see page 9).

3. Bake the white cake in a 9 × 13-inch cake pan.

4. Prepare the honey buttercream.

Making the Surprise

5. Using a 2-inch round cookie cutter, press down lightly on the center of one cake layer to create a guideline. Repeat on the other layer.

6. Using a small spoon, start carving around the guideline, digging a bit beyond the surface guideline to create the cavity that will be the white wing portion of the bee. Reserve the cutout cake in a bowl.

7. In the center of the hole you've made, make a well that's about 1½ inches across and 1½ inches deep to hold the yellow and black bee body.

8. Repeat the cuts with the second chocolate layer.

9. Tint 1 cup of the honey buttercream with 1 or 2 drops of yellow food coloring to make a yellow buttercream.

10. Make white and yellow cake mixtures (see page 11) using the white cake: Remove any dark or hard spots from the cake—it must be pure white. Then crumble half the cake into a large bowl and add 1 or 2 tablespoons white buttercream. Crumble the other half of the cake into a second bowl and add 1 or 2 tablespoons yellow buttercream. Combine each mixture until it achieves a Play-Doh-like consistency.

11. Make a brown cake mixture out of the reserved chocolate cake scraps mixed with 1 or 2 tablespoons of the buttercream.

12. Roll 1 cup of white cake mixture into a snake that's about 8 inches long and 1 inch in diameter. Repeat to make a separate snake.

13. Fit the white snake snugly inside the hollowed-out wing portion of the cake. Add more white cake mixture if you need to. The space that remains will hold the bottom half of the body of the bee, so you can see how big you'll need to make it. Repeat with the second snake and second cake layer.

14. Build matching bee-shaped bodies out of the yellow and brown cake mixtures.

16. To make the bee body wider at the center, flatten and widen the pieces a bit as you assemble it.

15. Carefully slice the bee bodies so that you can reconstruct a single yellow body with brown stripes.

17. Insert the bee body into the cake and build up around it with the white cake mixture.

18. When the area around the bee is filled in, remove the top half of the bee body and place it into the other cake layer. Build up the area around that body with more white cake mixture.

19. Tint ¾ cup of the buttercream with brown gel food coloring. Spread ½ cup in a very thin layer around the perimeter of one layer of the cake, avoiding the white cake mixture and the bee body. Gently flip the other layer on top so that the bee meets in the middle.

20. Cover the cake in a crumb coat of honey buttercream (see page 13) and chill for at least 1 hour.

Frosting and Decorating

21. Cover the cake in a smooth layer of honey buttercream (see page 12).

22. Insert a #2 decorating tip into a disposable pastry bag and fill it with the remaining ¼ cup of brown buttercream. Pipe out at least 3 dashed wavy lines on the outside of the cake to represent the path a bee might take as he travels up the side.

23. Put the chocolate chips in a disposable plastic pastry bag. Seal the bag with a rubber band and microwave it in 30-second intervals until the chips are melted. Cut a very small tip from the corner of the bag.

24. Working on a piece of parchment or wax paper on a baking sheet for stability, pipe the melted chocolate into the shape of a small bee (draw the bee on paper first if you want a pattern to follow). Add a line of chocolate below the bee; this will be the bee's stand. Make the line long enough so that you'll be able to insert a bit of it into the cake.

25. Place the chocolate bee in the freezer for 2 minutes, until firm.

26. Carefully peel away the paper and insert the stand into the cake where the piped wavy line leaves off.

27. Serve immediately.

cherry cake

SPECIAL EQUIPMENT:
*4 8-inch round cake pans;
9 × 13-inch pan; offset spatula;
4-inch round cookie cutter;
paring knife; rolling pin;
disposable plastic pastry bags;
coupler; #2, #3, #5, and
#7 decorating tips*

DIFFICULTY: *Challenging*

I've been on this cherry kick for about five years now, ever since our daughter Audrey came into our lives. There's just something about a little girl dolled up in a pretty dress covered in cherries that makes my heart happy.

About the only way I can actually *eat* a cherry is if it's chocolate-covered, but I thought I might be able to give it a go if it was in cake form. Although I simply used a white cake that was tinted red for my cherry, you can easily add a drop or two of cherry extract to your cake. Then you can actually bite into a cherry-flavored cake cherry!

Baking

1. Bake 4 white cake layers (from 2 recipes of white cake) in 8-inch round cake pans.

2. Prepare the batter for the last recipe of white cake, then add 4 to 6 drops of red gel food coloring, or enough to give you a bright red color. Bake the cake in a 9 × 13-inch pan.

3. Cool all the cakes to room temperature.

4. Make the buttercream.

5. You'll be making two 2-layer cakes out of the 4 white cake layers. Lay 1 white layer on a piece of parchment. Using an offset spatula, cover the top in a thin coat of buttercream (about ½ cup) and place another white layer on top. Make a second 2-layer cake with the 2 remaining white layers. Refrigerate the 2 cakes for at least 6 hours.

6. Remove any hard edges or brown spots from the red cake and crumble it into a large bowl. Add 2 to 4 tablespoons of buttercream and combine with a fork or your hands until the cake mixture is the consistency of Play-Doh. Refrigerate the cake mixture in a plastic storage bag or container.

7. Cover the side of the cakes in a crumb coat of buttercream (see page 13).

Making the Surprise

8. Using a 4-inch round cookie cutter, find the center of each cake and press in gently to create a guideline.

9. Use a small spoon (a baby spoon works great) to carve out a narrow circular channel (about 1 inch wide and 1 inch deep) in one of the white layer cakes.

10. Reserve the scraps of cake in a bowl.

11. Add 4 to 6 drops green food coloring to the white cake scraps in the bowl to form the cherry stem. Add buttercream, 1 teaspoon at a time, and combine until the cake mixture is the consistency of Play-Doh (see page 11).

12. Cutting on the guideline you made earlier on the second cake, insert your knife into the cake pointing toward the center at a 45-degree angle (insert a skewer into the center for guidance if you like). Cut around the entire cake and remove the cone shape. Set it aside.

13. Roll out the red cake mixture into small sections and fill the channel in the first cake, tucking it in snugly.

14. Place a handful of the green cake mixture on one side of a large piece of parchment. Fold the parchment over the cake mixture and roll it out to about ¼ inch thick.

15. Use the 4-inch cookie cutter to cut a circle out of the green cake mixture.

16. Cut a line halfway through the circle.

17. Turn the cutout cake cone point side up and drape the green circle over the cone. Wrap it around and make the green layer as smooth as possible; the cut will allow you to overlap the green. Then insert the green cone back into the well in the cake.

18. Spread a thin layer of buttercream on top of the cake with the cherry (just the white part). Gently flip over the green stem cake and lay it on top of the cherry cake (the cone should be facing up within the cake).

19. Refrigerate the cake for 1 hour or freeze it 15 minutes to set the design, then cover the cake with a crumb coat (see page 13).

20. Chill in the refrigerator until you're ready to decorate, but at least 1 hour.

Frosting and Decorating

21. Tint half of the buttercream turquoise (about 2 cups) and cover the cake in a smooth coat (see page 12).

22. Place a ½ cup of the remaining buttercream in a small bowl and tint it light green. Tint the remaining buttercream bright red.

23. Insert a coupler into a disposable plastic pastry bag and attach a #3 decorating tip. Fill the bag with the red buttercream. Pipe the outlines of small cherries around the cake. I used groupings of two cherries.

24. Insert a #2 decorating tip into a pastry bag and fill it with green buttercream. Pipe the stems of the cherries.

25. Change the tip on the bag of red buttercream to a #5 or #7 and fill in the cherries.

26. Chill for about 30 minutes, then smooth out any rough spots. I do this with my finger dipped in a little water (not too much!) or with a paper towel.

27. Change to a #5 tip and pipe red buttercream to create a border around the base of the cake. Pipe out a dot, then continue around the entire cake.

28. Chill the cake until serving.

military cake

Men in both my family and my husband's have made the amazing sacrifice of service to this country. While they served in different divisions, different countries, and different wars, the common denominator was that they were all army men. This cake is a small tribute to everyone who has served. To those who've chosen sacrifice over self, I'm in awe of you!

I decided to do stripes in this cake because they're one of the few symbols that are universally military. I have plenty to learn about what the different stripes and colors mean, but you can switch this cake up with whatever's appropriate for your favorite serviceman or servicewoman.

I'd be happy to join you in some research over a nice big piece of cake!

3 recipes White Cake, page 19

1 recipe Basic Buttercream, page 24

Green and brown gel food coloring

SPECIAL EQUIPMENT: *4 8-inch round cake pans; 9 × 13-inch pan; offset spatula; 4½-inch round cookie cutter; thin paring knife; skewer; rolling pin; serrated knife; disposable plastic pastry bags; #4 and #12 decorating tips; coupler set (optional)*

DIFFICULTY: *Challenging*

Baking

1. Use 2 recipes of white cake to bake four 8-inch round layers. Cool them to room temperature.

2. Bake the remaining recipe of white cake in a 9 × 13-inch pan and cool it to room temperature.

3. Make the buttercream.

4. Place one of the layers on a sheet of parchment and, using an offset spatula, spread a thin layer of buttercream on top. Place a second cake layer on top. Reserve the other 2 layers.

5. Freeze the layers for at least 6 hours, preferably overnight. There's no need to freeze the 9 × 13-inch cake.

Making the Surprise

6. Place a 4½-inch round cookie cutter in the center of the 2-layer cake and press as far as it will go without smushing the cake. Remove the cutter and use a sharp thin paring knife to finish the cut all the way to the bottom.

7. Carefully remove the center portion. I lifted the cake slightly and pressed the center portion up for easy removal.

8. Insert a skinny skewer into the center of the cake column. Using a sharp paring knife and starting from the outer edge, cut through the cake at a 45-degree angle until you reach the skewer. Cut around the entire perimeter of the cake column and remove the cone-shaped top. Lay the cone point side up.

9. Remove the hard bits and dark parts and crumble the 9 × 13-inch cake into 2 large bowls. To one bowl, add 2 to 4 drops of green gel food coloring and 1 or 2 tablespoons of buttercream. Mix very well to make a pliable cake mixture (see page 11). Add more food coloring as needed until the mixture is very green. Combine 1 or 2 tablespoons of buttercream to the other bowl of crumbled cake and combine to make a white cake mixture.

10. Place a fistful of green cake mixture on a piece of parchment and roll it out to a thickness of ½ inch. Cut out a green circle using the 4½-inch cookie cutter, then cut a line halfway through the circle.

12. Cut away the excess overlap and smooth out the green layer.

11. Place the circle on top of the cake cone you just made.

13. Repeat with a layer of white cake mixture on top of the green, followed by a second green layer.

14. Place one of the reserved layers on a cake stand and cover the top with a thin layer of buttercream.

15. Carefully set the 2-layer cake on top. Then slowly and gently place the stripe structure inside the cake, pointing up.

16. Invert the cut 2-layer column so that the cap fits over the green point of the stripe structure inside the cake. Using a serrated knife, slice off the excess cake to make an even top surface.

17. Frost the top with buttercream and place the final reserved layer on top.

18. Cover the cake in a crumb coat (see page 13) and chill for at least 1 hour.

Frosting and Decorating

19. Divide the remaining buttercream among 4 bowls. In 3 of the bowls, add different amounts of brown gel food coloring to achieve different shades of khaki and brown. Add green gel food coloring to the last bowl. Stir the 4 colors separately and add more color as needed to get the desired shades for camouflage.

20. Place the 4 colors in plastic disposable pastry bags. Use a coupler set to make it easier to switch out the colors while using the same decorating tip.

21. Insert a coupler into a pastry bag and attach a #4 decorating tip. Cut the tip off one of the frosting bags and slip it into the bag with the tip. Outline blob-like camouflage shapes around the cake. Switch colors until you are satisfied with the design.

22. Change to a #12 decorating tip to fill in the blobs with the matching color.

23. Freeze the finished cake for about 15 minutes, then use a plain paper towel to flatten the buttercream. Be gentle!

24. Chill the cake until serving.

rainbow heart cake

This cake is dedicated to all my beautiful, supportive, kind, and loving blogging friends. I feel as if I'd have given up on blogging years ago had they not given me the strength I needed to get through.

In the past three years I've seen many of them reach amazing successes . . . from TV appearances to books published to working with huge names like Martha Stewart, Bobby Flay, and Michelle Obama!

Food bloggers are unusually multitalented individuals. They can do everything from recipe development to stunning photography to high-level marketing. I'm constantly learning from these trendsetters and feel so blessed to call them friends!

3 recipes White Cake,
page 19

2 recipes Basic Buttercream,
page 24

Red, orange, yellow, green, blue, and purple gel food coloring

¼ cup Simple Syrup

SPECIAL EQUIPMENT:
4 8-inch cake pans; offset spatula; 6 6-inch cake pans; ruler; 3-inch round cookie cutter; thin paring knife; skewer; glazing brush; disposable plastic pastry bags; 1M decorating tip

DIFFICULTY: *Challenging*

simple syrup

Combine 1 cup sugar and 1 cup water in a saucepan and bring to a boil. Stir occasionally until the sugar dissolves, then cool the syrup to room temperature.

Baking

1. Bake 2 recipes of white cake into four 8-inch layers. Cool thoroughly.

2. Prepare the buttercream.

Making the Surprise

3. Using the offset spatula, cover the top of 1 layer in a thin coat of ¼ to ½ cup buttercream and top with another layer.

4. Repeat with the 2 remaining cake layers. You now have two 2-layer cakes.

5. Refrigerate the cakes for 4 hours to overnight or freeze them for at least 1 hour.

6. Make the batter for the third recipe of white cake. Divide the batter among 6 bowls, about ½ cup to ¾ cup batter per bowl.

7. Add 4 to 6 drops of food coloring to each bowl to make batter in each shade: red, orange, yellow, green, blue, and purple. (You can always modify the recipe to have 7 layers if you're a purist and want to see indigo in there!)

8. Bake each color in a 6-inch cake pan, working in batches depending on how many pans you have. Bake for just 10 to 12 minutes.

9. Set the colored cakes aside to cool, then remove them from the pans. Stack the layers with parchment between them. Start with red on the bottom: Weight can often flatten cakes a bit, so at this stage, stack them in the opposite order they'll appear in the cake.

10. Freeze the stack for about 30 minutes, or until firm but not fully frozen.

11. When you're ready to begin carving, remove the 2-layer cakes from the freezer.

12. Remove the rainbow layers from the freezer, turn them red layer up, and measure the stack to determine its height. The total height of my rainbow stack is about 4 inches. That means I need to carve down approximately 2 inches into each layer cake. Place the stack back in the refrigerator or freezer.

13. Place a 3-inch cookie cutter in the center of each 2-layer cake and gently press down enough to make a shallow guideline.

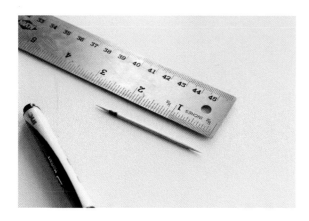

14. Start on the bottom of the cake (or the V part of the heart). Mark a toothpick or skewer at 2 inches and press it into the center of the cake just to the 2-inch mark.

15. Insert a thin, sharp knife at the guideline at about a 45-degree angle; you should feel the bottom of the toothpick with the knife. Cut around the perimeter of the circle.

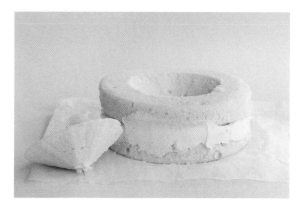

16. Remove the cone shape and put this layer cake back into the refrigerator or freezer.

17. Insert the marked toothpick or skewer into the center of the second 2-layer cake to the 2-inch mark. At a 45-degree angle, insert the knife at the guideline again, but almost vertically. You're trying to make a valley with a raised point in the center; this is the top of the heart. Move the knife in around the cake in one smooth motion, making sure your knife goes in only about 2 inches deep.

19. Using a soupspoon, carve around the channel to make a smooth, rounded top to the heart. Remember to carve out a bit of the "wall" of the heart, but stay within the guideline so that the top and bottom halves will match up nicely. If your cake feels too soft or crumbly, just place it in the freezer for a bit.

18. Next, insert the knife at a 45-degree angle into cake closer to the center of the cake. In one smooth motion, cut an even inner circle to make a channel, cutting only about 2 inches deep. Remove the section you've carved out. Put the layer cake back in the fridge or freezer if it's at all crumbly; it needs to be very chilled.

20. Remove the rainbow layers from the refrigerator and have ready the simple syrup and a glazing brush.

21. Place the red layer next to the cake with the channel cut from it. Eyeball where the center is and cut a circle from the red layer. Use the 4-inch cutter to trim around the edges to be sure it will fit the cavity of the cake.

22. Carefully place the red layer into the cavity. If it doesn't fit, remove it and make the necessary cuts and adjustments. A very chilled cake will allow you to handle it with less risk of breaking.

23. Once the red layer fits snugly, gently brush it with no more than 1 tablespoon of simple syrup.

24. Now set the orange cake layer next to the cake. Estimate how much you'll need to trim from around the edges and to remove from the center so that it fits snugly in the cavity. This can be a rough estimate—it doesn't need to be perfect. Make the cuts and adjustments, then gently place the orange layer into the cake and brush it with simple syrup.

25. Next, lay in the yellow layer, which will only need to be trimmed around the edges. Set this cake in the refrigerator or freezer and bring out the other 2-layer cake.

26. Lay out the purple layer. Roughly estimate the size of the bottom third of the cavity where the cone was removed; a small cone of purple cake goes there. Insert the knife at a 45-degree angle and cut out a small cone from the purple cake.

27. Insert the purple cone into the cavity of the white cake. If you've miscalculated the size of the purple cone, go back and make another—you have plenty of cake! Brush with a thin layer of simple syrup.

28. Repeat with the blue layer; making the circle slightly larger. Brush with a thin layer of simple syrup.

29. Finally, trim off the edges of the green layer so that it fits snugly on top of the blue layer.

30. Spread a thin layer of buttercream around the top perimeter of the white cake, avoiding the green cake.

31. Remove the other cake from the refrigerator, invert it, and carefully place it on top so that the yellow layer meets the green layer.

32. Freeze the cake for at least 2 hours. Don't skip this step! It's important to make sure those individual layers inside really bond together.

33. Cover the cake in a crumb coat (see page 13). Place the cake in the refrigerator for about an hour to set.

Frosting and Decorating

34. Cover the cake in a smooth coat of buttercream.

35. Decorate the cake in buttercream roses using the technique outlined for the Miniature Heart Cake (page 141).

36. Chill the cake until serving.

acknowledgments

My dear sweet adorable hottie husband, Chad—you are my rock. Thank you. Thank you for everything you do and are. I love you!

Colton, Parker, Audrey, Eddie, and Olivia, my precious children—you're my inspiration and joy. I love you tremendously! Remind Mommy of that next time you cut your sister's hair.

Cassie Jones—thanks for your help in translating these cakes onto the page. I could not appreciate you more! You and Kara Zauberman are wonderful! And thank you to the rest of the terrific team at William Morrow/HarperCollins: Liate Stehlik, Lynn Grady, Jennifer Hart, Megan Swartz, Ashley Marudas, Joyce Wong, Keonaona Peterson, Lorie Pagnozzi, and Karen Lumley.

Alison Fargis—you were my champion, even when I didn't deserve it. You do your job amazingly well, and I've loved getting to know you these past years. Stonesong rules!

Susan Powers—your talent is an inspiration: photographer, food stylist, and prop stylist all in one. Thank you for being a part of this, and thank you for your support. I look forward to always learning from you!

Kara Carlson and Londa Crigger—thank you! The input you provided in recipes has been priceless to me! You were brutally honest and thoughtfully supportive. I am grateful!

Ree Drummond—although you're far too humble to admit it, I know that without you none of this would have come to fruition. Thank you.

To the wonderful bloggers whom I look up to and am lucky enough to call friends: Cheryl Sousan, Kristen Doyle, Rose Atwater, Alice Currah, Julie Ruble, Grace Langlois, Carla and Matt Walker, Amy Atlas, Robyn Stone, Amy Johnson, Brenda Score, Shaina Olmanson, Stephanie Meyer, Naomi Robinson, Jenny Flake, Jen Shall, Heather Cristo, Susan Eaton, Amber Bracegirdle, Jamie Lothridge, Rachel Matthews, Rachel Currier, Amy Locurto, Amy Huntley, Shanna Schad, Carolyn Ketchum, Renee Goerger, Monique Volz, Lori Lange, Toni Dash, Erin Dee, Jaden Hair, Marla Meredith, Stephanie Alexander, Michael Lewicki, Susan Salzman, Aimee Wimbush-Bourque, Rachel Gurk, Courtney Lopez, Debi Dellicarpini, Dawn Finicane, Sally McKenney, Maria Lichty, Julie Deily, Cassie Laemmli, Katrina Bahl, Bree Hester, Sylvie Shirazi, Melissa Lanz, Amy Clark, Amy Bellgardt, Sweet Paul(s), Betsy Eves, Sarah De Heer, Melissa Diamond, Deborah Stauch, Tara Kuczykowski, and so many more! Thank you!

To all the faithful, loyal, and beyond supportive blog readers—I adore you. We have grown together and you have made me a better blogger! You are kind, encouraging, honest, and motivating. I look forward to getting to know you better and hopefully meeting you in person!

Dan and Deb McGinty (Dad and Mom)—thank you for offering to buy 100 copies so that at least 101 would be sold. You're my biggest cheerleaders! Dad, thanks for trying so hard to "get" the cakes!

Angie, Inga, and Kurt Fraser—thanks for letting me sit at Mom's table and go on and on about cake, and for sharing so many ideas. You are so supportive and sweet! I love you all and am so thankful for you!

Kirk and Shara Anderson—your support strengthened me. Your prayers mean so much. Your friendship has been invaluable. Thank you for being a constant source of Love and Truth.

Kim Schieches—thank you for being my biggest cheerleader. I am so grateful for your friendship. You are smart and funny and a beautiful person!

The fabulous folks at KitchenAid—the beautiful commercial mixer was put to the test. I made more than fifty cakes in thirty days and it never faltered! The team at KitchenAid has been great to work with, and I appreciate all the support!

To Grace, Cara, Michelle, Rob, and the rest of the Craftsy crew. I adore you and am in awe of you. Your talent and kindness are second to none! I learned more from you in three days than I ever knew possible.

To the outstanding team at Betty Crocker. You have been so wonderful to work with! Thank you for believing in me!

index

Note: Page references in *italics* indicate photos of finished cakes.